T0305684

Introduction to Simulations of Semiconductor Lasers

Simulations play an increasingly important role not only in scientific research but also in engineering developments. Introduction to Simulations of Semiconductor Lasers introduces senior undergraduates to the design of semiconductor lasers and their simulations. The book begins with explaining the physics and fundamental characteristics behind semiconductor lasers and their applications. It presumes little prior knowledge, such that only a familiarity with the basics of electromagnetism and quantum mechanics is required. The book transitions from textbook explanations, equations, and formulas to ready-to-run numeric codes that enable the visualization of concepts and simulation studies. Multiple chapters are supported by Matlab™ code which can be accessed by the students. These are ready-to-run, but they can be modified to simulate other structures if desired.

Providing a unified treatment of the fundamental principles and physics of semiconductors and semiconductor lasers, Introduction to Simulations of Semiconductor Lasers is an accessible, practical guide for advanced undergraduate students of Physics, particularly for courses in laser physics.

Key Features:

- A unified treatment of fundamental principles.
- Explanations of the fundamental physics of semiconductor.
- Explanations of the operation of semiconductor lasers.
- An historical overview of the subject.

Marek S. Wartak is a Professor in the Department of Physics and Computer Science at Wilfrid Laurier University, Waterloo, Ontario. He has over 30 years of experience in semiconductor physics, photonics and optoelectronics, analytical methods, modelling and computer-aided design tools.

Introduction to Simulations of Semiconductor Lasers

Marek S. Wartak

CRC Press
Taylor & Francis Group
Boca Raton London New York

CRC Press is an imprint of the
Taylor & Francis Group, an **informa** business

First edition published 2024
by CRC Press
2385 NW Executive Center Drive, Suite 320, Boca Raton FL 33431

and by CRC Press
4 Park Square, Milton Park, Abingdon, Oxon, OX14 4RN

CRC Press is an imprint of Taylor & Francis Group, LLC

© 2024 Marek S. Wartak

Library of Congress Cataloging-in-Publication Data

Names: Wartak, Marek S., author.
Title: Introduction to simulations of semiconductor lasers / Marek S. Wartak.
Description: First edition. | Boca Raton, FL : CRC Press, 2024. | Includes bibliographical references and index. | Summary: "Simulations play an increasingly important role not only in scientific research but also in engineering developments. Introduction to Simulations of Semiconductor Lasers introduces senior undergraduates to the design of semiconductor lasers and their simulations."
Identifiers: LCCN 2023043148 | ISBN 9781032209043 (hbk) | ISBN 9781032209050 (pbk) | ISBN 9781003265849 (ebk)
Subjects: LCSH: Semiconductor lasers--Computer simulation.
Classification: LCC QC689.55.S45 W37 2024 | DDC 621.36/6101133--dc23/eng/20231130
LC record available at https://lccn.loc.gov/2023043148

ISBN: 978-1-032-20904-3 (hbk)
ISBN: 978-1-032-20905-0 (pbk)
ISBN: 978-1-003-26584-9 (ebk)

DOI: 10.1201/9781003265849

Typeset in CMR10 font
by KnowledgeWorks Global Ltd.

Publisher's note: This book has been prepared from camera-ready copy provided by the authors.

Contents

Preface   xi

1 Introduction   1
- 1.1 Fundamentals of Lasers 2
    - 1.1.1 Transitions in a TLS 4
    - 1.1.2 Laser oscillations and resonant modes 5
- 1.2 Semiconductor Laser Diodes 7
    - 1.2.1 Types of semiconductor lasers 8
    - 1.2.2 Homogeneous p-n junction 11
    - 1.2.3 Heterostructures 12
    - 1.2.4 Basic characteristics 14
- 1.3 An Outline 15
- Bibliography 16

2 Fundamentals of Semiconductors   19
- 2.1 Crystal structure of semiconductors 19
- 2.2 Simplified Band Structure of Semiconductors 21
- 2.3 Equilibrium Behavior in Semiconductors 23
    - 2.3.1 Densities in semiconductors. Fermi-Dirac distribution function 23
    - 2.3.2 Degenerate and nondegenerate semiconductors 25
- 2.4 Doped Semiconductors 27
    - 2.4.1 Charge neutrality relations 29
- 2.5 Homostructures 30
    - 2.5.1 Energy band diagrams for homostructures 31
- 2.6 Heterostructures 32
    - 2.6.1 Energy bands in nonuniform semiconductors. Nondegenerate case. 32
    - 2.6.2 Abrupt heterostructure 34
    - 2.6.3 Interesting application 36
- 2.7 Double Heterostructure 37
- 2.8 Quantum well 38
- Bibliography 39

3 Semiconductor Transport Equations and Contacts   **41**
   3.1   Drift-Diffusion Model 41
   3.2   Concentrations in Non-Equilibrium Situations 45
       3.2.1   Boltzmann statistics 45
       3.2.2   Fermi-Dirac statistics 46
   3.3   Contacts 49
       3.3.1   Schottky barriers 50
       3.3.2   Ohmic contact 51
   3.4   Currents across Heterointerface 51
       3.4.1   Thermionic model 52
       3.4.2   Gradient model 55
   3.5   Appendix 57
   Bibliography 58

4 p-n Junctions   **61**
   4.1   Formation of p-n Homojunction 61
   4.2   Simple Model of Homojunction: Debye Length 64
       4.2.1   n-region 64
       4.2.2   p-region 65
   4.3   Homo Junction in the Depletion Approximation 66
       4.3.1   Mathematical Details of the 1D Model 66
   4.4   p-n Homojunction under Forward and Reverse Bias 70
   4.5   Model of p-n Junction with Ohmic Contacts 70
   4.6   p-i-n Diode 73
   4.7   Hetero p-n Junction 74
       4.7.1   Formation of heterojunctions 74
   Bibliography 78

5 Electrical Processes   **79**
   5.1   Basic Physical Constants 79
   5.2   Band Structure Parameters 80
       5.2.1   The effective densities of states 81
   5.3   Doping 81
   5.4   Carrier Mobilities 83
   5.5   Recombination 86
       5.5.1   Spontaneous recombination 87
       5.5.2   Stimulated recombination 87
       5.5.3   Shockley-Read-Hall (SRH) generation-recombination . 87
       5.5.4   Auger recombination 90
   Bibliography 93

6 Poisson Equation   **94**
   6.1   Simple Poisson Equation 95
   6.2   p-n Diode in Equilibrium 98
   6.3   Scaling of Poisson Equation 100

6.4   Boundary Conditions and Trial Values 101
    6.4.1   Boundary conditions for electrostatic potential 101
    6.4.2   Initial (trial) values for potential 103
6.5   Poisson Equation for Homojunction 103
    6.5.1   Method on: Contacts outside 103
    6.5.2   Method two: Contacts inside 106
6.6   Poisson Equation for Non-Uniform Systems 108
    6.6.1   Linearization 108
    6.6.2   Discretization 109
    6.6.3   Boundary conditions for potential 111
    6.6.4   Initial conditions for potential 111
6.7   Applications of Poisson Equation to Analyze p-n Diode 111
    6.7.1   General 111
    6.7.2   Analysis of convergence 111
    6.7.3   Homo-junction with linear doping 113
Bibliography 121

7   Experiments Using Poisson Equation: Homo diode   122
7.1   Method One 122
    7.1.1   Calculations of band edges 123
    7.1.2   Comments about mesh 123
    7.1.3   Description of functions 124
7.2   Method Two 126
7.3   Solution and Results 126
Bibliography 137

8   Hetero-Junction Using Poisson Equation   138
8.1   Heterostructure Diode with Step Doping 138
8.2   Summary of Implemented Equations 138
    8.2.1   Nonuniform system (heterostructure) 139
    8.2.2   Description of functions 139
    8.2.3   Results for homo-structure 142
    8.2.4   Data functions 144
    8.2.5   Calculations 146
    8.2.6   Test data 153
Bibliography 155

9   Homo-Diode Based on Drift-Diffusion   156
9.1   Electrical Equations 156
    9.1.1   SRH recombination 157
    9.1.2   Mobility models 157
    9.1.3   Boundary conditions 158
    9.1.4   Trial values 160
    9.1.5   Choice of electrical variables 161
    9.1.6   Summary of linearized Poisson equation 161

9.2 Integration of Current Continuity Equation 162
9.3 Approximations to Bernoulli Function 165
9.4 Steady State: Discretization 165
    9.4.1 Discretization of electrons and holes 166
9.5 Scaling 166
    9.5.1 Scaling at boundaries 168
    9.5.2 Scaling of trial values of potential 168
    9.5.3 Scaling of mobilities 169
    9.5.4 Scaling of recombination 169
    9.5.5 Scaling of continuity equations 169
9.6 Electric Current 170
9.7 Results 171
    9.7.1 Results at equilibrium 171
    9.7.2 Results for non-equilibrium 172
Bibliography 175

10 Matlab Code for p-n Homo-Diode   **177**
10.1 Summary of Implemented Equations: Homogeneous case . . . 177
    10.1.1 Main functions 178
    10.1.2 Definitions of parameters 195

11 Hetero-Diode Based on Drift-Diffusion   **197**
11.1 Poisson Equation in Equilibrium 197
11.2 Poisson Equation in Non-Equilibrium 198
11.3 Electrons 198
11.4 Holes 198
11.5 SRH Recombination 199
11.6 Currents 199
11.7 Parameters 199
    11.7.1 Mobilities 200
    11.7.2 Dielectric constant 200
11.8 Code Summary 200
11.9 Simulated Structures 200
11.10 Results 200
    11.10.1 Equilibrium case 201
    11.10.2 Non-equilibrium case 201
    11.10.3 Data files 205
    11.10.4 Extra functions 207
    11.10.5 Models 208
    11.10.6 Main files 210
Bibliography 225

12 Multi-Layer Passive Slab Waveguides   **226**
12.1   Modes of the Arbitrary Three Layer Asymmetric Planar Waveguide in 1D 226
12.2   Multilayer Waveguide 229
     12.2.1   Propagation matrix formulation 230
     12.2.2   Propagation constant 232
     12.2.3   Electric field 234
12.3   Testing 234
     12.3.1   6-layer lossy waveguide. 234
     12.3.2   p-i-n structure 235
12.4   List of Files 236
     12.4.1   Data files 236
     12.4.2   Extra files 240
     12.4.3   Main files 242
Bibliography 248

13 Optical Parameters and Processes   **250**
13.1   Optical Parameters 250
     13.1.1   Dielectric function and refractive index 250
     13.1.2   Static permittivity 252
     13.1.3   Optical gain 252
13.2   Absorption (losses) Coefficients 254
     13.2.1   Free-carrier absorption 255
     13.2.2   Intervalence band absorption 256
     13.2.3   The mirror loss 257
     13.2.4   Auger processes 257
13.3   Spontaneous emission factor 257
Bibliography 258

14 Semiconductor Laser   **261**
14.1   Summary of Electrical Equations 261
     14.1.1   Poisson equation in equilibrium 261
     14.1.2   Poisson equation in non-equilibrium 261
     14.1.3   Electrons 262
     14.1.4   Holes 262
14.2   Recombination Processes 262
     14.2.1   Recombination coefficients 264
14.3   Optical Equations 265
     14.3.1   Wave equation 265
     14.3.2   Photon rate equation 266
     14.3.3   Output power 268
     14.3.4   Practical photon rate equation 268
14.4   Remaining Material Parameters 268
     14.4.1   Static permittivity 268
     14.4.2   Carrier mobilities 268

14.5   Description of the Program 269
       14.5.1   Electrical part 269
       14.5.2   Optical part 269
       14.5.3   Full simulator 270
14.6   Results of Simulations 271
       14.6.1   Data files 273
       14.6.2   General 275
       14.6.3   Models 278
       14.6.4   Optical field 282
       14.6.5   Semiconductors 289
Bibliography 302

15 Conclusions   **303**
Bibliography 304

A Material Parameters   **305**
A.1   Some Properties of Important Materials 305
       A.1.1   Bandgap energies 305
       A.1.2   Mobilities 305
A.2   Practical Material: $Al_xGa_{1-x}As$ 307
       A.2.1   Band structure parameters 307
       A.2.2   Band discontinuity 307
       A.2.3   Doping 308
       A.2.4   Carrier mobilities 309
       A.2.5   Optical parameters 309
       A.2.6   Recombination parameters 310
       A.2.7   Losses 310
A.3 $In_{1-x}Ga_xAs_yP_{1-y}$ Material System 311
       A.3.1   Band discontinuity 311
       A.3.2   Doping 312
       A.3.3   Carrier mobilities 312
       A.3.4   Optical parameters 313
       A.3.5   Optical gain 313
       A.3.6   Recombination coefficients 314
       A.3.7   Absorption coefficients 315
       A.3.8   Spontaneous emission factor 316
       A.3.9   Summary of parameters for InP systems 316
Bibliography 317

B Short History of Semiconductor Laser Simulations   **320**
B.1   Companies 329
B.2   More Recent Developments 332
Bibliography 333

Index   **347**

Preface

Semiconductor lasers play an important role in today's technological developments. They can be found in many popular device, including iphones, DVD players, laser printers. Semiconductor laser (name laser diode is also often used) is based on p-n junction which is combined with an optical resonator. The p-n junction forms a diode which operates in a forward direction.

The aim of this book is a simple self contained introduction to the topic of semiconductor lasers with Matlab code and the examples of ready to run simulations. We advocate a simulation type approach to teach discussed topics. We provide a self-contained development which includes theoretical foundations and also Matlab code.

This book is aimed to be a very practical one with the following characteristics:

- provide complete theoretical background assuming only basic knowledge

- write and analyze computer code paing considerable attention to the proper values of physical parameters

- learn though computer simulations.

Our goal was to provide theoretical foundations and develop software framework skeleton for future experimentation. (By all means this is not a final (or commercial) software). I tried to make all chapters self-contained (to the extend possible), so some of the material is repeated.

The book contains 13 chapters and two appendices. It begins, in **Chapter 1** with explanation of laser principles and also as a general introduction to semiconductor lasers.

In **Chapter 2** we outline basic characteristics of semiconductors which are materials with resistivities having values intermediate between metals and insulators. These materials are also characterized by energy gap with a typical value of the order of 1eV.

In **Chapter 3** we summarized fundamental processes which undergo operation of semiconductor lasers involve the flow of electrons and holes and how they respond to all potentials (applied, build-in and scattering) and interaction of carriers with light. In that chapter, we concentrate on summarizing description of the flow of electrons and holes, known as carrier transport problem. Carrier transport in semiconductors has been extensively discussed by many authors.

In **Chapter 4** we describe the p-n junction which is the fundamental building block in electronics and optoelectronics. The junction found applications in many diodes, like: light emitting diodes (LED), semiconductor lasers, photodiodes, solar cells, Zener diodes. It can be formed within the same material by introducing different doping (such structure is known as homo p-n junction) or it can be formed between different materials (again doped) and such arrangement is termed as hetero-junction.

Fundamental electrical processes taking place in p-n junction are summarized in **Chapter 5**. They are determined by many material parameters. They are the most critical elements in determining quality of the simulations and directly affect simulation results.

The above chapters are establishing foundations for future developments. Next, we start with Poisson equation and take several steps of increasing complexity as the Poisson equation is central to the development of classical simulations of semiconductor lasers. And then we go over and include non-equilibrium cases which are based on drift-diffusion approach. In the entire book we restrict ourselves to the one-dimensional situations so a lot of problems are left outside of the scope of this book.

The simplest approach to model p-n junction is based on using Poisson equation, discussed in **Chapter 6**. Such approach describes equilibrium case. We use it to model homogeneous junction and also heterojunction.

Those experiments are described in **Chapter 7** where we report on using the previously developed methods to numerically solve Poisson equation for p-n diode. As another application of Poisson equation we discuss p-n junction with step doping in equilibrium.

In **Chapter 8**, developed methods are applied to numerically solve Poisson equation for hetero p-n diode. Hetero diode is formed by combining two materials with different properties, like bandgaps and effective masses. We describe simulations of two structures: hetero and homo based on AlGaAs as described in literature.

In **Chapter 9**, we advance model of p-n diode developed in the previous chapter by adding carrier transport equations to the Poisson equation already existing. Carrier transport of electrons and holes is described by drift-diffusion model. We also include fundamental recombination processes. Together they will provide realistic model of p-n junction and will be starting point for semiconductor laser model developed later. We concentrate on one-dimensional (1-D) case and also restrict ourselves to homogeneous material. Detailed Matlab code in provided in **Chapter 10**.

Based on the above developments, in **Chapter 11** we analyze heterostructure p-n junction.

In the following **Chapter 12**, we explain how propagation constant of electromagnetic wave is determined and used it to calculate profile of electric field. This part is independent from the rest of the simulator. It is also independently tested. Later-on it will be modified and integrated with the electrical part of the simulator. This chapter serves as a test of optical part of the simulator. It is not complete, for example we will not include dependency of refractive index on carrier density which is an important effect. We concentrate on one-dimensional situation and TE mode.

Optical processes specific to semiconductor lasers and also relevant parameters are summarized in **Chapter 13**. They include relative dielectric constant, refractive index and phenomenological parameters which describe optical gain. In the simplest, temperature and wavelength independent linear gain model is characterized by two parameters. In more complicated models, for example when temperature and/or wavelength dependencies are considered, there will be more parameters needed.

Finally, in **Chapter 14** we describe practical aspects of numerical implementation of the electrical and optical equations introduced in previous chapters. We concentrate on one-dimensional (1-D) case. Since some of the variables change by many orders of magnitude, it is a common practice to introduce appropriate scaling. After discussing it, other choice of variables will be discussed.

This book should be of interest to someone interested in computational electronics and photonics, and in particular semiconductor lasers. It contains Matlab code, ready for experimentation and possible improvements. It is our hope that young students and also more mature researchers alike will find something interesting and useful here. We welcome feedback for the future editions of this book.

The author would like to wish a potential reader the similar joy of reading this book and experimenting with programs as he had with writing it.

Finally, I welcome all type of comments from readers, especially concerning errors, inaccuracies and omissions. Please send your comments to *mwartak@wlu.ca*.

Requirements

I do assume that potential readers possess an undergraduate knowledge of electromagnetism and quantum mechanics. More specialized topics are covered. With it, one should be able to understand both, the discussed theory and Matlab code.

Acknowledgements

I want to thank my sister Agata Wartak for hospitality at her place where most of this book (or at least important parts of it) were written and I dedicate this book to her.

I also want to thank my former students and postdocs who contributed to my understanding. Thank you for your collaboration.

1

Introduction

In this introductory chapter, we outline the scope of this book by briefly describing semiconductor lasers. Consult Piprek's book [1] for a general source of information on simulations of optoelectronic devices and [2] for an introduction to the operating principles of laser diodes.

Over last twenty years or so, in addition to those mentioned above, several excellent books describing fundamentals and also different types of semiconductor lasers have been published or edited [3–27].

None of those books cover all aspects necessary to develop working simulator. We aim here to provide self-contained document which will cover not only fundamentals of physics of semiconductor lasers but also describe process of creating working simulator.

Semiconductor laser (name laser diode is also often used) is based on p-n junction, so, we start our discussion with required basis of semiconductors and then will cover operation of various types of p-n junctions.

An authorized history of semiconductor lasers has been provided by Alferov [28], and also personal account of how the double-heterostructure laser idea got started is provided by Kroemer [29].

According to Alferov, the idea of semiconductor laser is linked to John von Neumann [30] who suggested (in 1953) using the injection in Ge p-n junction for generation and amplification of infrared radiation.

The simplest laser diode is indeed just a p-n junction combined with an optical resonator. The p-n junction forms a diode which operates in a forward direction. In Fig. 1.1, we show the p-n junction and its electrical symbol, and in Fig. 1.2, we show the current-voltage characteristics of a p-n junction.

Actually, the modern semiconductor laser (laser diode) is much more complicated. Typical device consists of several layers of various semiconductors. Simplified view of the so-called Fabry-Perot (FP) semiconductor laser is shown in Fig. 1.3.

Next, we start with a more systematic description of operation of a laser.

DOI: 10.1201/9781003265849-1

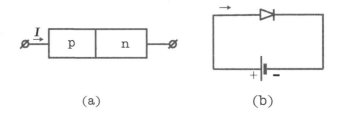

FIGURE 1.1
The simplest light source based on p-n junction formed with a single semiconductor, (a) p-n junction, and (b) its electrical symbol.

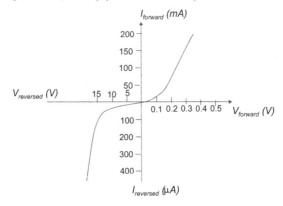

FIGURE 1.2
Current-voltage characteristics of p-n junction. Observe scale difference along vertical axis.

1.1 Fundamentals of Lasers

Generic laser structure is shown in Fig. 1.4 [31, 32]. It consists of a resonator (cavity), here formed by two mirrors, and a gain medium where the amplification of electromagnetic radiation (light) takes place. A laser is an oscillator analogous to an oscillator in electronics. To form an oscillator, an amplifier (where gain is created) and feedback are needed. Feedback is provided by two mirrors which also confine light. One of the mirrors is partially transmitting which allows the light to escape from the device. There must be an external energy provided into gain medium (process known as pumping). Most popular (practical) pumping mechanisms are by optical or electrical means.

Gain medium can be created in several ways. Conceptually, the simplest one is the collection of gas molecules. Such systems are known as gas lasers. In gas lasers, one can regard the active medium as an ensemble of absorption or amplification centers (e.g., like atoms or molecules) with only some electronic energy levels which couple to the resonant optical field. Other electronic states

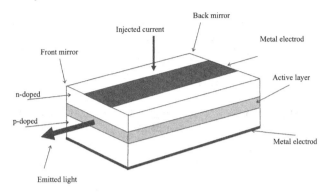

FIGURE 1.3
Simplified perspective view of FP semiconductor laser.

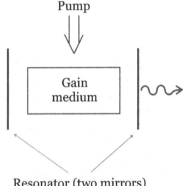

FIGURE 1.4
Generic laser structure: two mirrors with a gain medium in between. The two mirrors form a cavity, which confines the light and provides the optical feedback. One of the mirrors is partially transmitted and thus allows light to escape. The resulting laser light is directional, with a small spectral bandwidth.

are used to excite or pump the system. The pumping process excites these molecules into a higher energy level.

Out of many existing levels in a molecule, for the detailed analysis one typically selects two energy levels with values E_1 and E_2; see Fig. 1.5. As illustrated, three basic processes are possible: absorption, stimulated emission and spontaneous emission. Such system is known as a two-level system (TLS).

Such TLS are met very often in Nature. Generally, for an atomic system, as in the case under consideration, we can always separate just two energy levels, the upper level and the ground state, thereby forming TLS.

As mentioned above, electron can be excited into the upper level due to external interactions (for lasers, process known as pumping). Electrons can

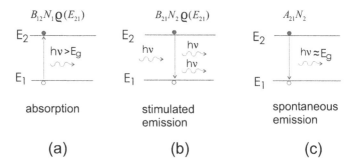

FIGURE 1.5
Illustration of possible transitions in two-level system and the relevant notation.

loose their energies radiatively (emitting photons) or non-radiatively, say by collisions with phonons.

For laser action to occur, the pumping process must produce population inversion meaning that there are more molecules in the excited state (here upper level with energy E_2) than in the ground state. If the population inversion is present in the cavity, incoming light can be amplified by the system; see Fig. 1.5 (b) where one incoming photon generates two photons at the output.

The way how TLS is practically utilized results in various types of lasers, like gaseous, solid state, semiconductor. Also, different types of resonators are possible which will be discussed in subsequent sections.

1.1.1   Transitions in a TLS

Assume the existence of two energy levels E_1 and E_2 (forming TLS), which are occupied with probabilities N_1 and N_2. Also introduce:

A_{21} the probability of spontaneous emission

B_{21} the probability of stimulated emission

B_{12} the probability of absorption.

The notation associated with those processes is illustrated in Fig. 1.5.

We introduce $\nu_{21} = (E_2 - E_1)/h$ and $\rho(\nu_{21})$ is the density of photons with frequency ν_{21}. Coefficients A_{21}, B_{21}, B_{12} are known as Einstein coefficients. The density of photons $\rho(\nu_{21})$ of frequency ν_{21} can be determined from the Planck's distribution of the energy density in the black body radiation as [32]

$$\rho(\nu_{21}) = \frac{8\pi h\nu_{21}^3}{c^3} \frac{1}{\exp\frac{h\nu_{21}}{kT} - 1} \tag{1.1}$$

For the laser we require that amplification be greater than absorption. Therefore, the number of stimulated transitions must be greater than the absorption transitions. Thus the net amplification can be created. In the following we will determine the condition for the net amplification in such systems [32].

The change in time of the occupancy of the upper level is

$$\frac{dN_2}{dt} = -A_{21}N_2 - B_{21}N_2\rho(\nu_{21}) + B_{12}N_1\rho(\nu_{21}) \tag{1.2}$$

First term on the RHS describes spontaneous emission, second term is responsible for stimulated emission and the last one for absorption. In thermal equilibrium which requires that $\frac{dN_2}{dt} = 0$, using Eq.(1.1), one obtains

$$N_2 \left\{ B_{21}\frac{8\pi h\nu_{21}^3}{c^3 \left(\exp\frac{h\nu_{21}}{kT} - 1\right)} + A_{21} \right\} = N_1 B_{12}\frac{8\pi h\nu_{21}^3}{c^3 \left(\exp\frac{h\nu_{21}}{kT} - 1\right)} \tag{1.3}$$

Next, assume that N_1 and N_2 are given by the Maxwell-Boltzmann statistics, i.e.

$$\frac{N_1}{N_2} = \exp\left(-\frac{h\nu_{21}}{kT}\right) \tag{1.4}$$

Substitute (1.4) into Eq.(1.3) and have

$$\exp\left(-\frac{h\nu_{21}}{kT}\right) \left\{ B_{21}\frac{8\pi h\nu_{21}^3}{c^3 \left(\exp\frac{h\nu_{21}}{kT} - 1\right)} + A_{21} \right\} = B_{12}\frac{8\pi h\nu_{21}^3}{c^3 \left(\exp\frac{h\nu_{21}}{kT} - 1\right)}$$

From the above equation, the density of photons $\rho(\nu_{21})$ can be expressed in terms of Einstein coefficients as

$$\frac{8\pi h\nu_{21}^3}{c^3 \left(\exp\frac{h\nu_{21}}{kT} - 1\right)} = \frac{A_{21}}{B_{12}\exp\left(\frac{h\nu_{21}}{kT}\right) - B_{21}} \tag{1.5}$$

In the above formula, both sides could be equal if

$$B_{21} = B_{12} \tag{1.6}$$

In such case, one finds

$$\frac{A_{21}}{B_{21}} = \frac{8\pi h\nu_{21}^3}{c^3} \tag{1.7}$$

The relations (1.6) and (1.7) are known as Einstein relations [32].

1.1.2 Laser oscillations and resonant modes

Light propagation with amplification is illustrated in Fig. 1.6. Mathematically it is described by assuming that there is no phase change on reflection at either end (left and right). Left end is defined as $z = 0$ and right end as $z = L$. At the right facet, the forward optical wave has a fraction r_R reflected (amplitude reflection) and after reflection that fraction travels back (from right to left).

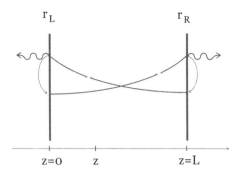

FIGURE 1.6
Schematic illustration of the amplification in a Fabry-Perot (FP) semiconductor laser with homogeneously distributed gain.

In order to form a stable resonance, the amplitude and phase of the wave after a single round trip must match the amplitude and phase of the starting wave. At an arbitrary point z inside the cavity (Fig. 1.6), the forward wave is

$$E_0 e^{gz} e^{-j\beta z} \tag{1.8}$$

where we have dropped $e^{i\omega t}$ term which is common and defined $g = g_m - \alpha_m$, where g_m describes gain (amplification) of the wave and α_m its losses. Also, r_R and r_L are, respectively, right and left reflectivities, L length of the cavity, and β propagation constant.

The wave travelling one full round will be

$$\left\{ E_0 e^{gz} e^{-j\beta z} \right\} \left\{ e^{g(L-z)} e^{-j\beta(L-z)} \right\} \left\{ r_R e^{gL} e^{-j\beta L} \right\}$$
$$\times \left\{ r_L e^{gz} e^{-j\beta z} \right\} \tag{1.9}$$

The above terms are interpreted as follows. In the first bracket there is an original forward propagating wave which started at z, in the second bracket there is wave travelling from z to L, the third bracket describes wave propagating from $z = L$ to $z = 0$, and the last one contains wave travelling from $z = 0$ to the starting point, z. At that point, the wave must match original wave as given by Eq.(1.8). From the above, one obtains condition for stable oscillations

$$r_R r_L e^{2gL} e^{-2j\beta L} = 1 \tag{1.10}$$

That condition can be split into amplitude condition

$$r_R r_L e^{2(g_m - \alpha_m)L} = 1 \tag{1.11}$$

and phase condition

$$e^{-2j\beta L} = 1 \tag{1.12}$$

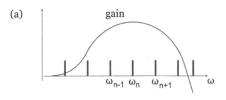

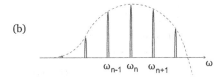

FIGURE 1.7
Gain spectrum of semiconductor laser and location of longitudinal modes are
the frequencies of FP resonances determined from phase condition.

From amplitude condition one obtains

$$g_m = \alpha_m + \frac{1}{2L} \ln \frac{1}{r_R r_L} \tag{1.13}$$

From phase condition, it follows that

$$2\beta L = 2\pi n \tag{1.14}$$

where n is an integer. The last equation determines the wavelengths of oscillations since

$$\beta = \frac{2\pi}{\lambda_n} = \frac{\omega_n}{c} \tag{1.15}$$

with λ_n being the wavelength. Typical gain spectrum and location of resonator
modes are shown in Fig. 1.7a. Longitudinal modes with angular frequencies
ω_{n-1}, ω_n and ω_{n+1} are shown. In time, the mode which has the largest gain
will survive, and the other modes will diminish, (see Fig. 1.7b).

1.2 Semiconductor Laser Diodes

Early developments of LED are described in Wikipedia [33]. Some important
results were reported in 1962 [34, 35] with gallium arsenide (GaAs) junction
operating at cryogenic temperatures. Those devices emitted light at around
850 nm. The wavelengths were subsequently extended into visible and near

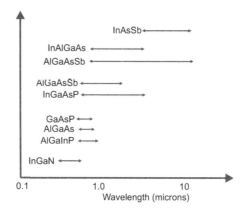

FIGURE 1.8
Emission wavelengths for some compounds and alloys.

infrared by employing other compound materials; see Fig. 1.8, which summarizes wavelength coverage using various materials from III-V groups. During this period of development, operating temperatures were also increased. Now the operation at room temperature is a standard.

More detailed account of the development of simulations of semiconductor lasers is provided in the Appendix.

Generally these light sources are divided into two groups:

(i) light-emitting diodes (LED), which emit incoherent light

(ii) semiconductor laser diodes (SLD), which are coherent emitters.

The above types of devices have similar structures. The main difference is associated with the existence of the so-called Fabry-Perot (FP) resonator cavity (or other ways to produce reflections) in the SLD. In those devices the emission of light is due to the radiative recombination of the electron-hole pairs (all these concepts will be explained later). Electron-hole pairs are formed by electrons from conduction band (CB) and hole from valence band (VB). Between these bands, there exists the forbidden band, characterized by bandgap energy.

1.2.1   Types of semiconductor lasers

As mentioned earlier, significant percentage of today's lasers are fabricated using semiconductor technology. Those devices are known as semiconductor lasers. Over the last fifteen years or so, several excellent books describing different aspects and different types of semiconductor lasers have been published [2, 3, 7, 9, 13].

The operation of semiconductor lasers as sources of electromagnetic radiation is based on the interaction between electromagnetic (EM) radiation and

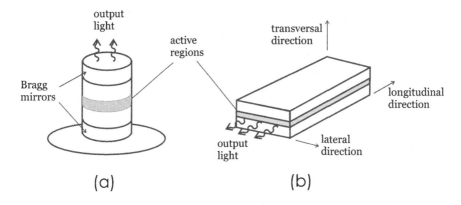

FIGURE 1.9
Comparison of VCSEL(left) and in-plane laser(right) semiconductor lasers.

the electrons and holes in semiconductor. Typical semiconductor laser structures are shown in Fig. 1.9. Those are: a vertical-cavity surface-emitting laser
(VCSEL) [36], where light propagates perpendicularly to the main plane, and
an in-plane laser where light propagates in the main plane [37]. The largest
dimension of in-plane structures is typically in the range of $250\mu m$ (longitudinal direction), whereas the typical diameter of a VCSEL cylinder is about
$10\mu m$.

The basic semiconductor laser is just a p-n junction; see Fig. 1.10, where
cross-section along lateral-transversal directions is shown. Current flows (holes
on the p-side and electrons on the n-side) along the transversal direction,
whereas light travels along the longitudinal direction and leaves device on one
or both sides.

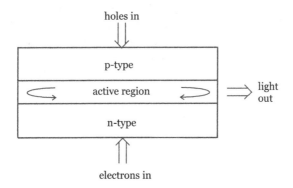

FIGURE 1.10
The basic p-n junction laser.

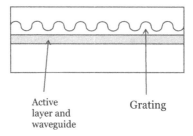

FIGURE 1.11
Basic DFB laser structure.

In VCSEL, the cavity is formed by the so-called Bragg mirrors, and an active region typically consists of several quantum well layers separated by barrier layers, (see Fig. 1.9). (For a detailed discussion of quantum wells see [3, 31]. This discussion is beyond the scope of this book). Bragg mirrors consist of several layers of different semiconductors which have different values of refractive indices. Due to the Bragg reflection, such structures show a very large reflectivity (around 99.9%). Such large values are needed because a very short distance of propagation of light does not allow to build enough amplification when propagating between distributed mirrors.

Three-dimensional perspective view of some generic semiconductor lasers was shown in Fig. 1.3. The structure consists of many layers of various materials, each engaged in a different role. Those layers are responsible for the efficient transport of electrons and holes from electrodes into an active region and for confinement of carriers and photons so they can strongly interact. Modern structures contain so-called quantum wells which form active region and where conduction-valance band transitions are taking place. It is possible to have different types of mirrors as they also provide mode selectivity. Two basic types are illustrated in Fig. 1.11 and Fig. 1.12 [37]. They are known as distributed feedback (DFB) and distributed Bragg (DBR) structures.

In DFB lasers, grating (corrugation) is produced in one of the cladding layers thus creating Bragg reflections at such periodic structure. The structure causes a wavelength sensitive feedback. It should be emphasized that the grating extends over entire laser structure. When one restricts corrugation to

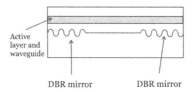

FIGURE 1.12
Basic DBR laser structure.

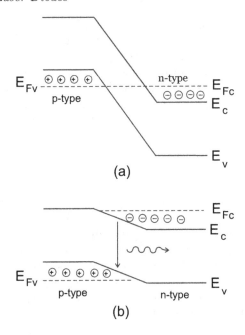

FIGURE 1.13
Energy-band diagram of a p-n junction: (a) thermal equilibrium, (b)forward
bias.

the mirror regions only and leaves flat active region in the middle, then so-
called DBR structure is created; see Fig. 1.12, which also provides wavelength
sensitive feedback.

1.2.2 Homogeneous p-n junction

Energy-band diagrams of a p-n junction in thermal equilibrium and under
forward bias conditions are shown in Fig. 1.13.

Electrons at the n-type side do not have enough energy to climb over
the potential barrier to the left. Similar situation exists for holes in the p-
type region (see Fig. 1.13a). One needs to apply forward voltage to lower the
potential barriers for both electrons and holes (see Fig. 1.13b).

An application of the forward voltage also separates Fermi levels. Thus,
two different so-called quasi-Fermi levels, E_{Fc} and E_{Fv}, are created, which
are connected with an external bias voltage as

$$E_{Fc} - E_{Fv} = eV_{bias} \qquad (1.16)$$

With the lowered potential barriers, electrons and holes can penetrate central
region where they can recombine and produce photons. However, the confinement of both electrons and holes into the central region is very poor (there

is no mechanism to confine those carriers). Also, there is no confinement of photons (light) into the region where electrons and holes recombine (the region is known as an active region). Therefore, the interaction between carriers and photons is weak, which makes homojunctions very poor light source. One must therefore provide some mechanism that will confine both carriers and photons into the same physical region where they will strongly interact. Such a concept was possible with the invention of heterostructures, which will be discussed next.

1.2.3   Heterostructures

Heterojunction is formed by joining dissimilar semiconductors. Basic type is formed by two heterojunctions and it is known as a double-heterojunction (more popular name is double-heterostructure).

Materials forming double heterostructure have different bandgap energies and different refractive indices. Therefore, in a natural way potential wells for both electrons and holes are created. Schematically, energy-band diagrams of a double-heterostructure p-n junction in thermal equilibrium and under forward bias conditions are shown in Fig. 1.14.

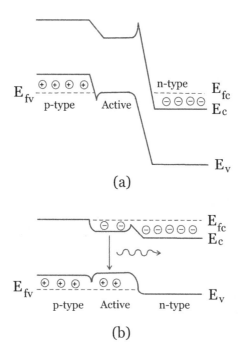

FIGURE 1.14
Energy-band diagram of a double-heterostructure p-n junction: (a) thermal equilibrium, (b) forward bias.

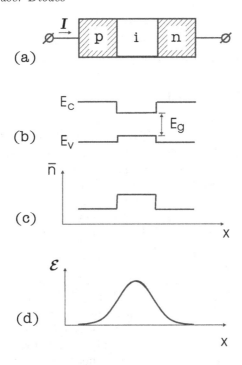

FIGURE 1.15
Main aspects of the double-heterostructure diode laser: (a) construction, (b) conduction and valence band edges, (c) form of refractive index, (d) profile of electric field.

Simplified vertical view of the laser structure fabricated based on double-heterostructure is shown in Fig. 1.15. Current is injected into a forward-biased p-n junction which produces population inversion. The structure consists of dissimilar semiconductor materials with the central region forming an active layer where the laser's action takes place. The regions which surround active region confine electromagnetic field. Along the longitudinal direction, the active layer forms a FP resonant cavity by two opposite reflecting faces, which are obtained by a cleavage of a single crystal.

For typical semiconductor materials used in optoelectronics, the refractive index at a given wavelength increases when the band-gap energy decreases. That fundamental material property creates larger value of refractive index in the active region which improves guiding of electromagnetic wave. The difference $\Delta \bar{n}$ in the values of refractive indices between guiding and barrier regions is typically very small. As a result of introduction of double heterostructure (DH), both electromagnetic (EM) field and carriers are confined into a very small region where they can interact very efficiently. The change in bandgap energies creates potential barriers for both holes and electrons. In the narrow region known as active region, the free charges recombine and create photons.

The active region has a higher refractive index (Fig. 1.15c) than the material on either side and forms efficient dielectric slab waveguide. The resulting profile of electric field is shown in Fig. 1.15(d).

1.2.4 Basic characteristics

Basic structures of semiconductor lasers are [37]: broad-area lasers, gain-guided lasers, weakly index-guided lasers, strongly index-guided lasers, surface emitting lasers. We will not discuss details of those structure here. We only show the cross section of ridge-waveguide structure (see Fig. 1.16), from [38].

The structure is formed by etching parts of the p-layer and depositing a SiO_2 layer, which blocks current flow. Refractive index of SiO_2 is lower than the central p-region. As a result, the effective index of the transverse mode is different in the two regions, which is enough to confine the generated light to the ridge region. This simple structure can be fabricated at a low cost, which makes it attractive for some applications.

Output power versus current, i.e. the so-called P-I characteristic, is shown in Fig. 1.17. No laser action is observed below threshold current. One can observe degradation of output power when temperature of the sample increases.

Far-field pattern emitted by a typical in-plane laser is shown in Fig. 1.18. Emitted radiation pattern determines coupling efficiency. The pattern is elliptical with a larger divergence in a plane perpendicular to the junction. Power diminishes as $cos\theta$, where θ is the angle between the viewing direction and the normal to the surface. The divergences parallel and perpendicular to the junction plane are shown in the figure and are typically $5-10°$ and $30-50°$, respectively.

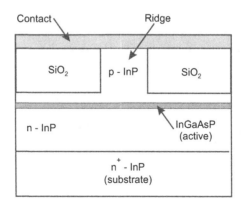

FIGURE 1.16
Cross section of index-guided ridge-waveguide structure used in weak index guiding semiconductor lasers.

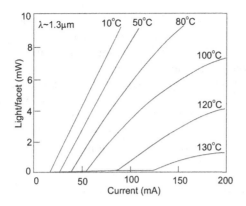

FIGURE 1.17
Output power of typical semiconductor laser.

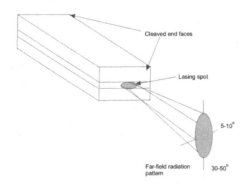

FIGURE 1.18
Far field patters from semiconductor laser.

Such characteristics make semiconductor lasers the most important type of lasers, technologically speaking. Their applications are extremely widespread and include optical telecommunications, optical data storage, metrology, spectroscopy, material processing, pumping of other lasers, and medical treatments.

1.3 An Outline

As I emphasized here, semiconductor lasers are fabricated using semiconductor materials (as the name suggests). Therefore in the following Chapter, I summarize the basic characteristics of semiconductors needed to understand the operation of lasers. After it, I will describe a basic p-n junction, as a semiconductor laser diode (SLD) is nothing more than a p-n junction.

In parallel, after establishing basic elements of p n junction I will develop Matlab code to simulate it. That code will be expanded to describe junction by more sophisticated techniques. And, finally we will provide details of semiconductor laser and develop Matlab code to describe it.

As the properties of optical modes are extremely important in the operation of SLD, we will also look at the waveguide properties.

Bibliography

[1] J. Piprek. *Semiconductor Optoelectronic Devices. Introduction to Physics and Simulations*. Academic Press, Amsterdam, 2003.

[2] D. Sands. *Diode Lasers*. Institute of Physics Publishing, Bristol and Philadelphia, 2005.

[3] B. Mroziewicz, M. Bugajski, and W. Nakwaski. *Physics of Semiconductor Lasers*. Polish Scientific Publishers, Warszawa, 1991.

[4] K. Iga. *Fundamentals of Laser Physics*. Plenum Press, New York, 1994.

[5] M.-C. Amann and J. Buus. *Tunable Laser Diodes*. Artech House, Boston, 1998.

[6] G. Morthier and PP. Vankwikelberge. *Handbook of Distributed Feedback Laser Diodes*. Artech House, Boston, 1997.

[7] L.A. Coldren and S.W. Corzine. *Diode Lasers and Photonic Integrated Circuits*. Wiley, New York, 1995.

[8] J.-M. Liu. *Photonic Devices*. Cambridge University Press, Cambridge, 2005.

[9] J. Carroll, J. Whiteaway, and D. Plumb. *Distributed Feedback Semiconductor Lasers*. The Institution of Electrical Engineers, SPIE Optical Engineering Press, United Kingdom, London, 1998.

[10] G. Guekos, editor. *Photonic Devices for Telecommunications*. Springer, Berlin, 1999.

[11] H. Ghafouri-Shiraz and B.S.L. Lo. *Distributed Feedback Laser Diodes*. Wiley, Chichester, 1996.

[12] P.S. Zory, Jr., editor. *Quantum Well Lasers*. Academic Press, Boston, 1993.

[13] K. Kapon, editor. *Semiconductor Lasers*. Academic Press, San Diego, 1999.

[14] J. Ohtsubo. *Semiconductor Lasers. Stability, Instability and Chaos*. Springer, Berlin, 2017.

[15] A. Baranov and E. Tournié, editors. Philadelphia, 2013. Woodhead Publishing.

[16] T. Numai. *Fundamentals of Semiconductor Lasers*. Springer, New York, 2015.

[17] T. Suhara. *Semiconductor Laser Fundamentals*. Marcel Dekker, New York, 2004.

[18] P.K. Basu, B. Mukhopadhyay, and R. Basu. *Semiconductor Laser Theory*. CRC Press, Boca Raton, 2016.

[19] T. Cunzhu and J. Chennupati, editors. Amsterdam, 2019. Elsevier.

[20] D.S. Patil, editor. Rijeka, Croatia, 2012. InTech.

[21] D.J. Klotzkin. *Introduction to Semiconductor Lasers for Optical Communications. 2nd Edition*. Springer Nature, Switzerland AG, 2020.

[22] A.F.J. Levi. *Essential Semiconductor Laser Device Physics*. Morgan and Claypool Publishers, San Rafael, California, 2018.

[23] A. Baranov and E. Tournie, editors. *Principles of Semiconductor Lasers*. Elsevier, 2013.

[24] R. Herrick and O. Ueda, editors. Elsevier, 2021.

[25] D. Lenstra, editor. MDPI, Basel, 2020.

[26] T. Cunzhu and C. Jagadish, editors. Elsevier, 2019.

[27] P.B. Bisht. *An Introduction to Photonics and Laser Physics with Applications*. IOP Publishing Ltd, Bristol, UK, 2022.

[28] Z.I. Alferov. The history of heterostructure lasers. In M. Grundmann, editor, *Nano-Optoelectronics. Concepts, Physics and Devices*, pages 3–22. Springer, 2002.

[29] H. Kroemer. Special 30th anniversary. how the double-heterostructure laser idea got started. In *IEEE LEOS Newsletter, August*, volume 21, pages 4–6. 2007.

[30] J. von Neumann. Collected works. volume 5, page 420. Pergamon Press, Oxford, 1963.

[31] A. Yariv. *Quantum Electronics. Third Edition*. Wiley, New York, 1989.

[32] C.C. Davis. *Lasers and Electro-Optics. Fundamentals and Engineering.* Cambridge University Press, Cambridge, 2002.

[33] https://*en.wikipedia.org/wiki/Light − emitting_diode*

[34] R.N. Hall, G.E. Fenner, J.D. Kingsley, T.J. Soltys, and R.O. Carlson. Coherent light emission from GaAs junctions. *Phys. Rev. Lett.*, 9:366–368, 1962.

[35] N. Holonyak and S.F. Bevacqua. Coherent (visible) light emission from $Ga(As_{1-x}P_x)$ junctions. *Appl. Phys. Let.*, 1:82–83, 1962.

[36] C.W. Wilmsen, H. Temkin, and L.A. Coldren, editors, *Verical-Cavity Surface-Emitting-Lasers.* Cambridge University Press, Cambridge, 1999.

[37] G.P. Agrawal and N.K. Dutta. *Semiconductor Lasers. Second Edition.* Kluwer Academic Publishers, Boston, 2000.

[38] G.E. Agrawal. *Fiber-Optic Communication Systems. Third Edition.* Wiley, New York, 2002.

2

Fundamentals of Semiconductors

2.1 Crystal structure of semiconductors

In order to establish fundamentals for readers with no experience with semiconductor electronics we provide a summary of basic properties of semiconductors.

Semiconductors are materials with resistivities having values intermediate between metals and insulators. In a semiconductor, a typical value is $100\Omega \cdot cm$, as compared to the value of the copper $10^{-6}\Omega \cdot cm$.

These materials are also characterized by energy gap. (We will talk about energy gap later). Typical value for a semiconductor is of the order of $1eV$ whereas for an insulator it is around $5eV$.

The most popular semiconductors are silicon (Si) and germanium (Ge). They belong to the group-IV elements in the periodic table and crystallize in a diamond structure. Basic element of diamond structure is a tetrahedron shown in Fig. 2.1. Here carbon, C atom, is at the center, and its four nearest-neighbors (NN) are at the corners of the cube. Each atom forms four bonds with its NN. Atoms in diamond-type crystals form covalent bonding.

The 3D tetrahedral structure is often represented in a simplified form on a flat surface, as shown in Fig. 2.2. Each circle represents the core of a semiconductor atom formed with electrons in inner orbits, and each line represents a shared valence electrons. This picture shows how a missing atom, point defect, or broken atom–atom bond could free an electron.

Important semiconductors are combinations of two, or more materials, forming so-called compounds. Popular semiconductor compound, gallium-arsenate (GaAs), is a III-V semiconductor. It crystalizes in the zincblende structure which is a variation of the diamond structure. The crystal structure of GaAs is shown in Fig. 2.3.

DOI: 10.1201/9781003265849-2

19

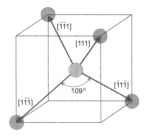

FIGURE 2.1
Tetrahedron structure.

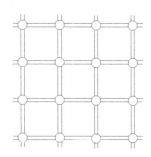

FIGURE 2.2
Bonding model.

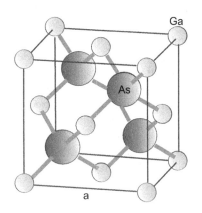

FIGURE 2.3
Crystal structure of GaAs.

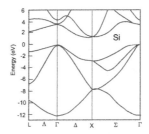

FIGURE 2.4
Band structure of Si.

2.2 Simplified Band Structure of Semiconductors

Popular methods used to determine band structure of semiconductors are [1]:

- The tight-binding (TB) method uses the atomic orbitals as basis for the wave function.

- The augmented-plane wave method was developed by Slater in 1937.

- The pseudopotential method differs from the above discussed methods by the manner in which the wave function is chosen.

Nonlocal pseudopotential calculations for the electronic structure of Si and GaAs are shown in Figs.2.4 and 2.5, after [2]. We illustrate two important cases, namely indirect (e.g. Si) and direct (GaAs) semiconductors, where transitions between VB and CB can be vertical on not, as shown in Fig. 2.6.

In optoelectronic applications (e.g. semiconductor lasers) central role is played by semiconductors with direct band transitions. In such materials near the minimum or maximum of the conduction or valence bands, respectively, bands can be approximated by parabolas, as shown in Fig. 2.7. One can thus

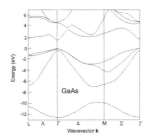

FIGURE 2.5
Band structure of GaAs.

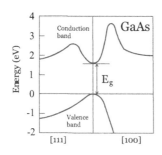
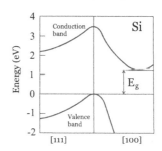

FIGURE 2.6
Illustration of direct and indirect bandgaps in GaAs and Si.

write

$$E(k) = E_g + \frac{\hbar^2 k^2}{2m_c^*}, \quad \text{conduction band} \tag{2.1}$$

$$E(k) = -\frac{\hbar^2 k^2}{2m_v^*}, \quad \text{valence band} \tag{2.2}$$

Here m_c^* and m_v^* are effective masses of carriers within conduction and valence bands, respectively. Reference energy is at the top of the valence band and E_g is the energy bandgap. Those are known as parabolic bands. The effect of non-parabolicity in conduction band is typically accounted as

$$E(k)\,(1 + \alpha E(k)) = E_g + \frac{\hbar^2 k^2}{2m_c^*} \tag{2.3}$$

where α is the coefficient of non-parabolicity. The above is the second-order

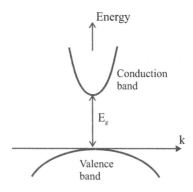

FIGURE 2.7
Illustration of parabolic bands.

algebraic equation which has the following solution

$$E(k) = \frac{1}{2\alpha}\left[\sqrt{1 + 4\alpha\left(E_g + \frac{\hbar^2 k^2}{2m_c^*}\right)} - 1\right] \tag{2.4}$$

2.3 Equilibrium Behavior in Semiconductors

In this section we concentrate on equilibrium properties of semiconductors. By definition, in equilibrium there is no net current flow of any kind. Flow of current (flux) will be discussed later.

We start with summarizing basic facts needed to describe semiconductors. Following Sze [3], we consider three basic types of a semiconductors (more about doping of semiconductors will be discussed shortly):

(a) *intrinsic semiconductor,* which contains a negligible small amount of impurities. Each atom shares its four valence electrons with the four neighboring atoms forming four covalent bonds,

(b) *n-type,* where atoms from group V (e.g. phosphorus) are added as impurities, thus 'donating' electrons to the conduction band,

(c) *p-type,* where atoms from group III (e.g. boron) are added as impurities, thus 'accepting' electrons and creating a positive charged hole in the valence band.

2.3.1 Densities in semiconductors. Fermi-Dirac distribution function

Electrons in semiconductors obey Fermi-Dirac (F-D) statistics, which says that in thermal equilibrium their distribution is [4]

$$f(E) = \left(1 + \exp\frac{E - E_F}{k_B T}\right)^{-1} \tag{2.5}$$

where k_B is Boltzmann's constant ($k_B = 8.62 \times 10^{-5} eV/K = 1.38 \times 10^{-23} J/K$). The function $f(E)$, the Fermi-Dirac distribution function, gives the probability that an available energy state at energy E will be occupied by an electron at absolute temperature T. The quantity E_F is called the Fermi energy (level). The occupation probability for an energy E equal to E_F is $f(E_F) = (1 + 1)^{-1} = \frac{1}{2}$, which tells us that the energy state at the Fermi level has a probability of $\frac{1}{2}$ of being occupied by an electron. At temperature $T = 0$, all the electron's states with energies $E \leq E_F$ are occupied.

In the limit of high energies, i.e. $E \gg E_F$ the exponent is very large and distribution function takes the form

$$f(E) \approx e^{-\frac{E-E_F}{k_B T}}$$

which is known as the Boltzmann distribution. For energies that are much lower than E_F the F-D distribution attains a constant value. We will look at this problem in more detail in the next section.

The location of the Fermi level can be changed by external parameters, like doping, current, etc. If the Fermi level is positioned in the conduction band it means that there are electrons filling the conduction band. Such materials conduct electric current very well. All metals belong to this category.

Fermi-Dirac distribution function is used to determine the concentrations of electrons and holes in semiconductors. (Concept of holes and electrons will be discussed in the next section in the connection of doping semiconductors.)

Assuming parabolic band model for electrons in the conduction band, one obtains the following expression for concentration of electrons at equilibrium [4]

$$n = \int_{E_c}^{E_{top}} f(E) g_c(E) dE \tag{2.6}$$

where E_c is the energy at the bottom of the conduction band and E_{top} is the energy at the top. $g_c(E)$ is the density of states which at the bottom of the conduction band is approximated as [4]

$$g_c(E) = \frac{m_n^* \sqrt{2m_n^*(E - E_c)}}{\pi^2 \hbar^3} \tag{2.7}$$

with m_n^* being the density-of-state effective mass for electrons. Letting $E_{top} \to \infty$ which is justified by the fact that the integrand falls off rapidly with increasing energy, one can evaluate integral in Eq.(2.6) with Fermi-Dirac distribution (2.5) and obtain for electron concentration

$$n = N_c F_{1/2} (\eta_c) \tag{2.8}$$

where

$$\eta_c = \frac{E_F - E_c}{k_B T} \tag{2.9}$$

and N_c is the effective density of states in the conduction band and is given by

$$N_c = 2 \left(\frac{m_n^* k_B T}{2\pi \hbar^2} \right)^{3/2} \tag{2.10}$$

Fermi-Dirac integral $F_{1/2}(\eta)$ of order one-half appearing in Eq. (2.8) is defined as [5]

$$F_{1/2}(\eta) = \frac{2}{\sqrt{\pi}} \int_0^\infty \frac{\sqrt{y}}{1 + \exp(y - \eta)} dy \qquad (2.11)$$

As shown, for example by Shur [5], the limiting forms of the function $F_{1/2}(\eta)$ are

$$\text{for } \eta \ll -1, \quad F_{1/2}(\eta) \approx \exp(\eta) \qquad (2.12)$$

and

$$\text{for } \eta \gg 1, \quad F_{1/2}(\eta) \approx \frac{4\eta^{3/2}}{3\sqrt{\pi}} \qquad (2.13)$$

Other approximation for the Fermi-Dirac integral is [6]

$$F_j(\eta) = \frac{1}{e^{-\eta} + C_j(\eta)} \qquad (2.14)$$

For $j = \frac{1}{2}$, we have (max error is 0.5%)

$$C_{1/2}(\eta) = \frac{3\pi^{1/2}/4}{\left(\eta^4 + 33.6\eta \left\{1 - 0.68 \exp\left[-0.17\left(\eta + 1\right)^2\right]\right\} + 50\right)^{3/8}} \qquad (2.15)$$

Properties and numerical approximations of Fermi-Dirac integrals are extensively discussed in [7–9], and [10].

Analogously, for concentration of holes near the top of the valence band

$$p = N_v F_{1/2}\left(\eta_v\right) \qquad (2.16)$$

where

$$\eta_v = \frac{E_v - E_F}{k_B T} \qquad (2.17)$$

Here E_v is the energy at the top of the valence band and N_v is the effective density of states in the valence band and is given by

$$N_v = 2 \left(\frac{m_p^* k_B T}{2\pi^2 \hbar^2}\right)^{3/2} \qquad (2.18)$$

where m_p^* is the density of states effective mass of the valence band.

2.3.2 Degenerate and nondegenerate semiconductors

Semiconductors are further classified [4] as degenerate and nondegenerate. In practice, degenerate semiconductors have Fermi level E_F which lies in the band-gap closer than $3k_B T$ to either conduction or valence bands, whereas in nondegenerate semiconductors E_F is in the region $E_v + 3k_B T \leq E_F \leq$

$E_c - 3k_BT$. Using dimensionless variables η_c and η_v the requirements for a semiconductor to be nondegenerate are

$$\eta_c \leq -3 \tag{2.19}$$

$$\eta_v \leq -3 \tag{2.20}$$

In the nondegenerate case, i.e. for $\eta_c \ll 1$ (in other words, for $F_F \ll E_c$) the unity in the denominator can be neglected, thus reducing the Fermi function to the Boltzmann distribution [9]

$$F_{1/2}(\eta) \approx e^{-\eta} \tag{2.21}$$

In practice it means that the Fermi level is several k_BT below E_c. In the nondegenerate limit, the Fermi-Dirac integral approaches $e^{-\eta}$ and the expression for electron concentration n, Eq.(2.8) becomes

$$n = N_c e^{\eta_c} = N_c \exp\left(\frac{E_F - E_c}{k_BT}\right) \tag{2.22}$$

For arbitrary values of η_c approximation techniques must be used to determine concentration of electrons. Similarly, under nondegenerate conditions, concentration of holes is expressed by the Boltzmann distribution

$$p = N_v e^{\eta_v} = N_v \exp\left(\frac{E_v - E_F}{k_BT}\right) \tag{2.23}$$

Concentrations of nondegenerate semiconductors are thus described by Eqs. (2.22) and (2.23).

Alternative formulas for densities can be introduced by recalling that E_i is the Fermi level for an intrinsic semiconductor [4]. Setting $n = p = n_{int}$ and $E_i = E_F$, one obtains

$$n_{int} = N_c \exp\left(\frac{E_i - E_c}{k_BT}\right) \tag{2.24}$$

and

$$n_{int} = N_v \exp\left(\frac{E_v - E_i}{k_BT}\right) \tag{2.25}$$

Solving for N_c and N_v from the above equations, gives

$$N_c = n_{int} \exp\left(\frac{E_c - E_i}{k_BT}\right) \tag{2.26}$$

and

$$N_v = n_{int} \exp\left(\frac{E_i - E_v}{k_BT}\right) \tag{2.27}$$

Substituting the last results into Eqs. (2.22) and (2.23), we obtain

$$n = n_{int} \exp\left(\frac{E_F - E_i}{k_BT}\right) \tag{2.28}$$

$$p = n_{int} \exp \left(\frac{E_i - E_F}{k_B T} \right) \tag{2.29}$$

From above equations, one finds

$$n \cdot p = n_{int}^2 \tag{2.30}$$

which is true for nondegenerate semiconductor in thermal equilibrium. By taking products of Eqs.(2.24) and (2.25), we obtain another usefull expression

$$n_{int} = \sqrt{N_c N_v} \exp \left(-\frac{E_c - E_v}{2k_B T} \right) = \sqrt{N_c N_v} \exp \left(-\frac{E_g}{2k_B T} \right) \tag{2.31}$$

where E_g is the bandgap energy defined as $E_g = E_c - E_v$.

2.4 Doped Semiconductors

Carrier concentration in semiconductors can be increased and controlled by doping using elements from group III or V. With doping, only small energies are required to produce free carriers. In the following we discuss separately two types of doping.

Using bonding model we can represent incorporation of an element from group V (say Arsenic, As) which has five valence electrons, as shown in Fig. 2.8.

Arsenic donates one electron to the crystal lattice. Such replacement of the original group IV atoms by atom from V group is known as n-type doping. It generates free electrons but it also creates the same number of positive ions, so charge neutrality is preserved. As an electron is donated, therefore such doping is known as *donors*.

One can also replace original atoms from group IV by atoms from group III, as shown in Fig. 2.8. In that case one bond will be 'free' so local positive charge will be created. Such arrangement is known as a p-type doping and

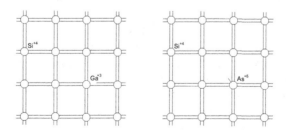

FIGURE 2.8
Doping in the bonding model. Replacing Si atom by Ga 'removes' one electron, thus creating a hole, whereas replacing Si by As adds one extra electron.

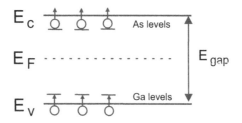

FIGURE 2.9
Doped semiconductor. Donor and acceptor levels for typical semiconductors are shown in the bandgap.

dopands are known as *acceptors*. The (local) objects created are known as holes.

Impurities, defects and dopants result in energy levels within the bandgap. Typical location of donor and acceptor states within energy gap is shown in Fig. 2.9.

For doping states the ratio of filled to unfilled states is expressed in terms of Fermi level as [18]

$$\frac{N_A^-}{N_A} = \frac{1}{g_A} \exp\left(\frac{E_F - E_A}{k_B T}\right) \tag{2.32}$$

and

$$\frac{N_D}{N_D^+} = g_A \exp\left(\frac{E_F - E_D}{k_B T}\right) \tag{2.33}$$

where N_D^+ is the number of ionized (positively charged) donors/cm^3 and N_A^- is the number of ionized (negatively charged) acceptors/cm^3. In semiconductors at room temperature, there is enough thermal energy to ionize almost all of the shallow level donors and acceptors. Assuming total ionization of dopant atoms, one has $N_D^+ = N_D$ and $N_A^- = N_A$. In the following, we will assume total ionization.

Here, E_A and E_D are the energies of acceptor and donor levels, and g_A is the ground state degeneracy factor for acceptor levels equal to 4 for Si, Ge and GaAs, because each acceptor level is doubly degenerate and can accept one hole of either spin. Here g_D is the ground state degeneracy factor of the donor impurity level equal to 2. In Table 2.1 we summarized values of binding energies for some semiconductors [4].

Relation for electron density becomes $n = N_D$, where N_D is the density of donors. One assumes here that essentially all free electrons originate from donors. Electron density therefore is

$$n = N_D = N_c \exp\left(\frac{E_F - E_c}{k_B T}\right) \tag{2.34}$$

TABLE 2.1

Dopant binding energies.

Donors	E_B	Acceptors	E_B
Sb	$0.039eV$	B	$0.045eV$
P	$0.045eV$	Al	$0.067eV$
As	$0.054eV$	Ga	$0.072eV$
		In	$0.16eV$

Similarly, for p-type (holes dominate) one has

$$p = N_A = N_v \exp\left(\frac{E_v - E_F}{k_B T}\right) \tag{2.35}$$

2.4.1 Charge neutrality relations

General condition for local charge neutrality requires that local charge is zero. It can be expressed as

$$p - n + N_D - N_A = 0 \tag{2.36}$$

Assuming nondegeneracy, we can also write (using Eq.(2.30))

$$p = \frac{n_{int}^2}{n} \tag{2.37}$$

Combining the above equations, we obtain algebraic equation used to determine concentration of electrons n

$$n^2 - n(N_D - N_A) - n_{int}^2 = 0 \tag{2.38}$$

Solving the above quadratic equation, one finds the following expression for n

$$n = \frac{N_D - N_A}{2} + \left[\left(\frac{N_D - N_A}{2}\right)^2 + n_{int}^2\right]^{1/2} \tag{2.39}$$

Similarly, we obtain the expression for concentration of holes p

$$p = \frac{N_A - N_D}{2} + \left[\left(\frac{N_A - N_D}{2}\right)^2 + n_{int}^2\right]^{1/2} \tag{2.40}$$

Carrier's concentrations must be positive or zero, so only plus signs were retained in the above relations. Important special cases which result from above equations are [4]:

Intrinsic semiconductor when $\mathbf{N}_A = 0$ and $\mathbf{N}_D = 0$.

In such situation, from the above equations, one obtains $n = n_{int}$ and $p = n_{int}$.

Doped semiconductor when $\mathbf{N}_D - \mathbf{N}_A \simeq \mathbf{N}_D \gg n_{int}$.

This is an important practical case. In this limit we obtain for concentrations of electrons and holes

$$\begin{aligned} n &\approx N_D \\ p &\approx \frac{n_{int}^2}{N_D} \end{aligned}$$

Doped semiconductor when $\mathbf{N}_A - \mathbf{N}_D \simeq \mathbf{N}_A \gg n_{int}$.

In this limit, we obtain for concentrations of electrons and holes

$$\begin{aligned} p &\approx N_A \\ n &\approx \frac{n_{int}^2}{N_A} \end{aligned}$$

Doped semiconductor when $n_{int} \gg |\mathbf{N}_D - \mathbf{N}_A|$.

This happens at sufficiently large temperatures. One finds

$$n \simeq p \simeq n_{int}$$

This means that all semiconductors become intrinsic at sufficiently high temperatures where $n_{int} \gg |\mathbf{N}_D - \mathbf{N}_A|$.

Compensated semiconductor when $\mathbf{N}_D - \mathbf{N}_A \simeq 0$, i.e. they are comparable and nonzero. Their opposing electrical effects are partially canceled. In this case both $\mathbf{N}_D$ and $\mathbf{N}_A$ must be retained in all expressions for carrier's concentrations.

2.5 Homostructures

An electrically pump semiconductor laser has the basic structure of a semiconductor diode. It typically operates under forward bias. Theoretically this device can be build utilizing either a homostructure or a heterostructure. All practical semiconductor lasers are fabricated based on heterostructures because of significant improvement of carrier confinement and optical waveguiding with respect to homostructures.

Usually, the term *homostructure* is used in the context of a p-n junction and refers to the same material (in particular bandgap remains the same over junction region). In contrast, *heterostructure* is associated with at least two different materials having two different bandgaps.

In what follows we will analyze operation of homo- and hetero-structures forming p-n junction and their usefulness as semiconductor lasers.

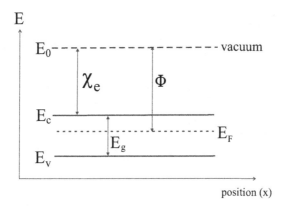

FIGURE 2.10
Energy band diagram of a uniform semiconductor (homostructure).

2.5.1 Energy band diagrams for homostructures

Understanding of transport in semiconductors is directly linked to its energy band diagram. For compositionally uniform semiconductors with uniform doping the energy band diagram looks like in Fig. 2.10 [11].

Conduction band edge (E_c) and valence band edge (E_v) energies characterize locations of conduction and valence bands, respectively. They can be measured or calculated. These energies are determined with respect to some reference energy (E_0, vacuum level). The electron affinity (χ_e) determines location of E_c with respect to vacuum. Work function (Φ) is the difference between vacuum energy and the Fermi energy. Location of the valence band edge energy (E_v) is determined by the energy band gap (E_g) with respect to E_c. In summary, for homostructures the electron affinity and bandgap are position independent and one has the following relations [12]

$$E_c = E_0 - \chi_e \tag{2.41}$$

and

$$E_v = E_c - E_g = E_0 - \chi_e - E_g \tag{2.42}$$

When slowly varying electric field $\mathbf{E}(x)$ is present in a semiconductor, the electron's energy changes as it moves under the field $\mathbf{E}(x)$ as

$$\Delta E = -\int F_e dx = -q\psi(x)$$

where F_e is the force acting on electron and $\psi(x)$ is the electrostatic potential. In such situation, the relations (2.41) and (2.42) must be modified as

$$E_c(x) = E_0 - \chi_e - q\psi(x)$$

and

$$E_v(x) = E_0 - \chi_e - E_g - q\psi(x)$$

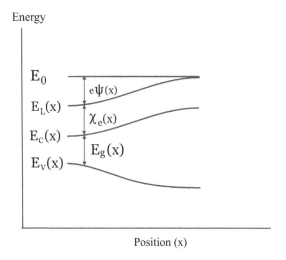

FIGURE 2.11
Energy versus position for a semiconductor with nonuniform composition.

The resulting energy band diagram shown in Fig. 2.10 will not be constant in space. In Fig. 2.11 we illustrated the situation for semiconductors with nonuniform composition and nonuniform electric potential.

2.6   Heterostructures

Semiconductor heterostructures play central role in semiconductor lasers. One forms a semiconductor heterostructure by placing two different semiconductor materials into contact. In practice this is done by growing different semiconductor on top of another semiconductor. Since semiconductors forming heterostructure are of different types, many of their properties are distinctly different. The most important of those properties are: lattice constants, energy gaps and electron affinities.

2.6.1   Energy bands in nonuniform semiconductors. Nondegenerate case.

In this Section we concentrate our discussion on the nondegenerate case described by Boltzmann statistics. In the degenerate case one must use Fermi-Dirac statistics.

The energy band diagram of a semiconductor with nonuniform composition is shown in Fig. (2.11) (from [12]). Here, E_0 is a reference level (vacuum level), $E_L(x)$ the local vacuum level, $\psi(x)$ is the local electrostatic potential, $\chi_e(x)$ local electron affinity.

The following equations hold (they are slightly more general then Eqs.(2.41) and (2.42)).

$$E_L(x) = E_0 - q\psi(x) \tag{2.43}$$

$$E_c(x) = E_0 - \chi_e(x) - q\psi(x) \tag{2.44}$$

$$E_v(x) = E_c(x) - E_g(x) \tag{2.45}$$

In the following, we will need an expression for an intrinsic energy E_i for heterostructures. To derive it we use previous expressions for p and n in terms of N_c and N_v (eqs. (2.22) and (2.23)) and also expressions (2.28) and (2.29). One finds

$$N_c \exp\left(\frac{E_F - E_c}{k_B T}\right) = n_{in} \exp\left(\frac{E_F - E_i}{k_B T}\right) \tag{2.46}$$

and

$$N_v \exp\left(\frac{E_v - E_F}{k_B T}\right) = n_{in} \exp\left(\frac{E_i - E_F}{k_B T}\right) \tag{2.47}$$

From Eqs.(2.46) and (2.47) one obtains useful expression for N_c and N_v valid for nondegenerate semiconductors

$$N_c = n_{in} \exp\left(\frac{E_c - E_i}{k_B T}\right) \tag{2.48}$$

and

$$N_v = n_{in} \exp\left(\frac{E_i - E_v}{k_B T}\right) \tag{2.49}$$

Divide Eqs.(2.46) and (2.47) by sides and obtain

$$\frac{N_c}{N_v} \exp\left(\frac{-2E_F + E_v + E_c}{k_B T}\right) = \exp\left(\frac{2(E_i - E_F)}{k_B T}\right)$$

Next, take logarithm of the above. After some simple algebraic manipulations with the help of equations (2.44) and (2.45), one finds the intrinsic energy level [12]

$$E_i(x) = E_0 - \chi_e(x) - q\psi(x) - \frac{1}{2}E_g(x) + \frac{k_B T}{2} \ln\frac{N_v}{N_c} \tag{2.50}$$

Remember that the above expression is true in the nondegenerate limit where Boltzmann statistics applies. One can observe that, in general $E_i(x)$ is not parallel to $\psi(x)$. Similar expression can be derived in the non-equilibrium case, when one assumes the existence of two quasi-Fermi energies.

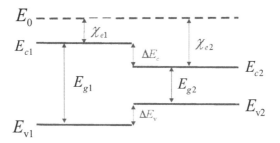

FIGURE 2.12
Abrupt heterointerface.

2.6.2 Abrupt heterostructure

Assume the two materials forming heterostructure have energy gaps E_{g1} and E_{g2}. Since the energy gaps are different, the conduction and valence bands of both materials cannot simultaneously be continuous across heterointerface. Generally, both conduction band and valence band edges are discontinuous at a heterointerface. The energy differences between the conduction band and valence band edges at the interface are called the conduction band and valence band discontinuities, respectively.

Terminology for abrupt heterointerface is illustrated in Fig. 2.12 (also known as type I heterostructure). Here E_{c1} and E_{c2} are conduction band edge energies, E_{v1} and E_{v2} are valence band edge energies and χ_{e1} and χ_{e2} are electron affinities of both semiconductors.

From Fig. 2.12, we observe that

$$\Delta E_c = \Delta \chi_e = \chi_{e2} - \chi_{e1} \tag{2.51}$$

Equation (2.51) is known as the electron affinity rule. The rule asserts that the conduction band offset at an abrupt heterojunction is equal to the difference in electron affinities between the two semiconductors. The semiconductor with the smaller affinity has higher conduction band at the interface. For these type of heterojunctions, we also have

$$\Delta E_g = \Delta E_c + \Delta E_v \tag{2.52}$$

where $\Delta E_g = E_{g1} - E_{g2}$ is the difference in the bandgaps (assuming $E_{g1} > E_{g2}$).

The above simple rule does not work well for a real semiconductors. When two semiconcuctors are brought together, there is a charge transfer across the interface due to the dissimilar nature of the chemical bonds. As a result, a short ranged dipole created at the interface modifies the band offsets.

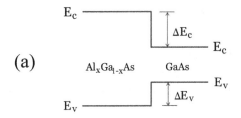

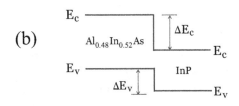

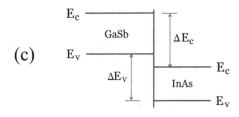

FIGURE 2.13
The three types of heterostructures: (a) type I, (b) type II, (c) type III.

Various possible bandedge lineups in heterostructures are shown in Fig. 2.13 [4, 13, 14]. They are known as type I, type II and type III. The relative position of conduction and valence bands is determined by the electron affinity. Examples of practical materials forming such configurations are presented below. Each of these types has a unique device application. In technological applications type I structure is mostly used. The three types of heterostructures are:

Type I heterostructure

In type I structure (straddeled band lineup) the conduction band edge and valence band edge of the material with smaller band gap are located within bandgap of the material with larger bandgap. Typical example is *AlGaAs – GaAs*. In type I heterostructure, the sum of the conduction band and valence band edge discontinuities is equal to the energy gap difference

$$\Delta E_g = \Delta E_c + \Delta E_v$$

In this structure electrons and holes will localize in the lower bandgap material.

Type II heterostructure

In this situation the discontinuities have different signs. In types II and III structures a staggered lineups are created. In those structures electrons and holes will localize in different materials.

Type III heterostructure

Here band structure is such that the top of the valence band in one material lies above the conduction band minimum of the other material.

Summary of important terminology:

1. Φ - work function. $e\Phi$ is the energy required to move an electron from the Fermi level to the vacuum level. Process which corresponds to removing an electron from the material.

2. χ_e (or χ) - electron affinity. $e\chi$ is the energy required to move an electron from the conduction band edge to the vacuum level.

3. ΔE_c and ΔE_v - are the conduction band and valence band discontinuities.

For a practical material system $Al_x Ga_{1-x} As$ and $GaAs$ the conduction and valence bands of the smaller bandgap semiconductor lie completely within the bandgap of the wider bandgap semiconductor. For example, for Al composition $x = 0.3$, the values of important parameters are $E_{g1} = 1.80eV$, $E_{g2} = 1.42eV$, $\Delta E_c = 0.23eV$ and $\Delta E_v = 0.15eV$.

2.6.3 Interesting application

Interesting application of type II heterostructure based on work reported by Khurgin et al [15] is now briefly discussed. The idea of using type II quantum well active region was to provide a practical way to increase a (usually) short lifetime of free-carriers. The considered structure is shown in Fig. 2.14. We have illustrated this situation for one period of the active region for $AlGaInAs/AlGaAsSb/AlInAs$ (from [15]). In that structure, electron wave functions are concentrated in the $2.5nm$ well fabricated from $Al_{0.2}Ga_{0.3}In_{0.5}As$, whereas the holes are confinement in the $3nm$ $Al_{0.5}Ga_{0.5}As_{0.5}Sb_{0.5}$ layer. The lasing wavelength of this structure is $1.25\mu m$. Authors of ref. [15] evaluated the transparency density which was $1.8 \times 10^{18} cm^{-3}$. The evaluated recombination lifetime was $11ns$, limited mostly by Auger and radiative processes. The value is about 10 times longer than in

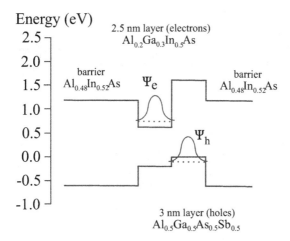

FIGURE 2.14
Band diagram for one period of the active region of the type-II, from [15].

type I structures operating in the same wavelength region. More details about application and operation of this structure are provided in [15].

2.7 Double Heterostructure

Double heterostructures (DH) played critical role in the development of semiconductor lasers. They are formed when three materials with different band gaps are brought together. $AlGaAs - GaAs - AlGaAs$ is a typical practical structure. In such a structure, significant confinement of carriers in space is possible to achieve. Also, larger value of refractive index of $GaAs$ compared to $AlGaAs$ regions facilitates the formation of efficient optical waveguide where electric field interacts with electrical carriers.

For working designs of semiconductor lasers the material with lower band separation is sandwiched between materials with larger band separation, thus forming double heterostructure. In those structures typical thickness of the inner layer is around $0.1\mu m$, therefore quantum effects are not significant. In Fig. 2.15 we show schematically energy band diagram of a typical NpP double heterostructure based on $AlGaAs/GaAs$ materials [16].

Cases without and with applied bias voltage are shown. In the situation shown, potential wells for both electrons and holes are formed which is evident from the Figure. This is achieved by the appropriate choice of doping and applied voltage.

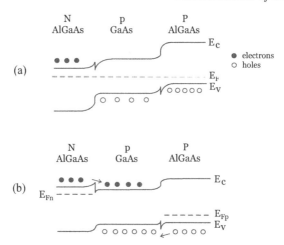

FIGURE 2.15
A double heterojunction system of NpP structure. Energy band diagram without (a) and with (b) applied bias voltage are shown.

2.8   Quantum well

To improve properties of DH semiconductor lasers a quantum well (one or more) is added and forms active region. It is just a very thin layer of semiconductor. Typical thickness is in the range of $5 - 10nm$. From conceptual point of view it is just a very thin double heterostructure. Because of its thickness, the effects of quantum confinement of electrons and holes starts to show up in the direction perpendicular to the junction plane and are responsible for the improved characteristics of the resulting devices.

The advantages over bulk samples are due to the fact that the injected carriers are concentrated in quantized subbands of the quantum well. As a result, the structure containing quantum well has a much larger gain than a bulk semiconductor. Also much lower injection current density is required for a quantum well structure than that for double heterostructure to reach transparency. Those structures also show higher modulation bandwidth. Their properties can be further improved by incorporating strain inside quantum well.

We finish this Section with emphasizing the role played by so-called separate confinement heterostructures (SCH), as shown in Fig. 2.16. They are introduced to improve optical waveguiding properties. Quantum wells are very thin and therefore show very poor waveguiding properties. Different shapes of SCH are possible. Those are linearly, parabolic and graded. Waveguiding properties of such structures were widely discussed in literature [17, 19].

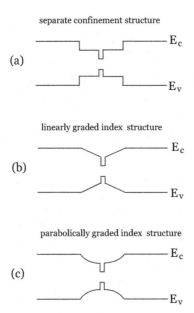

FIGURE 2.16
Different configurations of separate confinement (SCH) heterostructures with single quantum well; (a) step, (b) linearly and (c) parabolically graded index structures.

Bibliography

[1] N.W. Ashcroft and N.D. Mermin. *Solid State Physics.* Holt-Saunders, New York, 1976.

[2] James R. Chelikowsky and Marvin L. Cohen. Nonlocal pseudopotential calculations for the electronic structure of eleven diamond and zinc-blende semiconductors. *Phys. Rev.* **B**, 14:556–582, 1976.

[3] S.M. Sze. *Physics of Semiconductor Devices.* Wiley, New York, 1969.

[4] R.A. Pierret. *Semiconductor Device Fundamentals.* Addison-Wesley, Reading, Massachusetts, 1996.

[5] M. Shur. *Physics of Semiconductor Devices.* Prentice Hall, Englewood Cliffs, NJ, 1990.

[6] S.-L. Chuang. *Physics of Photonics Devices.* Wiley, New York, 2009.

[7] J. McDougall and E.C. Stoner. The computation of Fermi-Dirac functions. *Phil. Trans. Roy. Soc. London,* 237:67–104, 1938.

[8] J.S. Blakemore. Approximations for Fermi-Dirac integrals, especially the function $F_{1/2}(\eta)$ used to describe electron density in a semiconductor. *Solid State Electron.*, 25:1067–1076, 1982.

[9] L.A. Coldren and S.W. Corzine. *Diode Lasers and Photonic Integrated Circuits.* Wiley, New York, 1995.

[10] Shen S. Li. *Semiconductor Physical Electronics.* Plenum Press, New York, 1973.

[11] B. Mroziewicz, M. Bugajski, and W. Nakwaski. *Physics of Semiconductor Lasers.* Polish Scientific Publishers, Warszawa, 1991.

[12] M.S. Lundstrom and R.J. Schuelke. Modeling semiconductor heterojunctions in equilibrium. *Solid State Electron.*, 25:683–691, 1982.

[13] J. Singh. *Physics of Semiconductors and Their Heterostructures.* McGraw, 1993.

[14] K.F. Brennan. *The Physics of Semiconductors.* Cambridge University Press, Cambridge, 1999.

[15] J.B. Khurgin, I. Vurgaftman, and J.R. Meyer. Modeling of Q-switched semiconductor lasers based on type-II quantum wells: Increasing the pulse energy and peak power. *Appl. Phys. Let.*, 80:2631–2633, 2002.

[16] K.J. Ebeling. *Integrated Optoelectronics. Waveguide Optics, Photonics, Semiconductors.* Springer-Verlag, Berlin, 1993.

[17] M. Yamada, S. Ogita, T. Miyabo, and Y. Nashida. A theoretical analysis of lasing gain and threshold current in GaAs-AlGaAs SCH lasers. *Trans. IECE Japan*, E69:948–955, 1986.

[18] K. Yang, J. R. East, and G.I. Haddad. Numerical modeling of abrupt heterojunctions using a thermionic-field emission boundary condition. *Solid State Electronics*, 36:321–330, 1993.

[19] S.R. Chinn, P.S. Zory, and A.R. Reisinger. A model for GRIN-SCH-SQW diode lasers. *IEEE J. Quantum Electron.*, 24:2191–2214, 1988.

3

Semiconductor Transport Equations and Contacts

Fundamental processes which undergo operation of semiconductor lasers involve the flow of electrons and holes and how they respond to all potentials (applied, build-in and scattering), interaction of carriers with light (which consists of photons) and propagation of electromagnetic wave (light). In the present chapter we concentrate on summarizing description of the flow of electrons and holes, known as carrier transport problem. Carrier transport in semiconductors has been extensively discussed by many authors (see classical review [1], and also more recent one [2]).

In many of those systems carrier transport is associated with so-called junction region (p-n junctions are discussed in the next chapter). The junction in semiconductors usually means the transition region between two regions having different electrical properties. Those may be associated, for example with doping, or maybe something else. In the following sections we will discuss transport of carriers in both situations, i.e. when there is a uniform region, and when heterostructure is involved.

3.1 Drift-Diffusion Model

Under non-equilibrium conditions (which can be created for example by applying high electric field to a semiconductor sample), the electron and hole concentrations will differ from their equilibrium values and they can no longer be represented by a single quantity E_F. At this stage one usually introduces two new energy parameters E_{Fn} and E_{Fp} called, respectively, the quasi-Fermi levels for electrons and holes. They are related to the corresponding quasi-Fermi potentials ϕ_{Fn} and ϕ_{Fp} as $E_{Fn} = q\phi_{Fn}$ and $E_{Fp} = q\phi_{Fp}$. The non-equilibrium carrier concentrations are thus expressed as [3]

$$n(\mathbf{r}) = N_c F_{1/2}\left(\sigma_n(\mathbf{r})\right) \tag{3.1}$$

$$p(\mathbf{r}) = N_v F_{1/2}\left(\sigma_p(\mathbf{r})\right) \tag{3.2}$$

DOI: 10.1201/9781003265849-3

where $\sigma_n(\mathbf{r})$ and $\sigma_p(\mathbf{r})$ are sometimes called Planck potentials [4]. We introduce separate notation here for Planck potentials to avoid conflict with Eqs.(2.9) and (2.17) (here we deal with two (quasi) Fermi levels instead of one). They are position dependent quantities and in three-dimensional model depend on a position vector $\mathbf{r}$. Explicitely, they are defined as

$$\sigma_n(\mathbf{r}) = \frac{E_{Fn}(\mathbf{r}) - E_c(\mathbf{r})}{k_B T} \tag{3.3}$$

$$\sigma_p(\mathbf{r}) = \frac{E_v(\mathbf{r}) - E_{Fp}(\mathbf{r})}{k_B T} \tag{3.4}$$

In the limit of Boltzmann statistics, one can approximate carrier concentrations as

$$n(\mathbf{r}) = N_c \exp\left(\sigma_n(\mathbf{r})\right) \tag{3.5}$$

$$p(\mathbf{r}) = N_v \exp\left(\sigma_p(\mathbf{r})\right) \tag{3.6}$$

Static and dynamic behaviour of carriers is described by carrier continuity equations and the current-density relations. Here we illustrate them in 3D. Continuity equations for electrons and holes are [3]

$$\frac{\partial n}{\partial t} = \frac{1}{q} \nabla \cdot \mathbf{J}_n(\mathbf{r}) - R + G \tag{3.7}$$

$$\frac{\partial p}{\partial t} = -\frac{1}{q} \nabla \cdot \mathbf{J}_p(\mathbf{r}) - R + G \tag{3.8}$$

where n the electron concentration, p the hole concentration, $\mathbf{J}_n$ is the electron current density, $\mathbf{J}_p$ is the hole current density, R describes recombination processes and G incorporates generation phenomena, such as impact ionization or carrier generation by external radiation.

The current densities of electrons and holes are given by drift-diffusion expressions which are

$$\mathbf{J}_n(\mathbf{r}) = q D_n \nabla n(\mathbf{r}) - q \mu_n n(\mathbf{r}) \nabla \psi(\mathbf{r}) \tag{3.9}$$

$$\mathbf{J}_p(\mathbf{r}) = -q D_p \nabla p(\mathbf{r}) - q \mu_p p(\mathbf{r}) \nabla \psi(\mathbf{r}) \tag{3.10}$$

where and D_n and D_p are electron and hole diffusion constants, and μ_n and μ_p are electron and hole mobilities, respectively, and $\psi(\mathbf{r})$ is the potential.

Under nondegenerate conditions Einstein relations relate the diffusion coefficients and mobilities as [5]

$$D_n = \mu_n \frac{k_B T}{q}, \qquad D_p = \mu_p \frac{k_B T}{q} \tag{3.11}$$

where k_B is Boltzmann's constant and T carrier's temperature. Electric field $\mathbf{E}$ is related to field quantities as

$$\mathbf{E}(\mathbf{r}) = -\nabla \psi(\mathbf{r}) = \frac{1}{q} \nabla E_c(\mathbf{r}) = \frac{1}{q} \nabla E_v(\mathbf{r}) \tag{3.12}$$

where $\psi(\mathbf{r}), E_c(\mathbf{r}), E_v(\mathbf{r})$ are illustrated in Fig. 2.11.

There exist alternative expressions for current densities where they are expressed in terms of quasi-Fermi energies. We will now derive alternative expression for $\mathbf{J}_n$. For this, assume low electric fields and nondegenerate semiconductors (Boltzmann statistics). Electron concentration is thus given by the equation (3.5). Substitute that equation into expression for current (3.9) and have

$$\mathbf{J}_n(\mathbf{r}) = qD_n\nabla\left\{N_c\exp\left(\frac{E_{Fn}(\mathbf{r}) - E_c(\mathbf{r})}{k_BT}\right)\right\} - q\mu_n n\nabla\psi(\mathbf{r}) \qquad (3.13)$$

Evaluate derivative

$$\nabla\left\{N_c\exp\left(\frac{E_{Fn}(\mathbf{r}) - E_c(\mathbf{r})}{k_BT}\right)\right\} = n(\mathbf{r})\left\{\frac{1}{k_BT}\left[\nabla E_{Fn}(\mathbf{r}) - \nabla E_c(\mathbf{r})\right]\right\}$$

Substitute the last result into equation for current $\mathbf{J}_n$ (3.13), use (3.12) and have

$$\begin{aligned}
\mathbf{J}_n(\mathbf{r}) &= qD_n\frac{1}{k_BT}n(\mathbf{r})\left\{\nabla E_{Fn}(\mathbf{r}) - \nabla E_c(\mathbf{r})\right\} + q\mu_n n(\mathbf{r})\frac{1}{q}\nabla E_c(\mathbf{r}) \\
&= \frac{qD_n}{k_BT}n(\mathbf{r})\nabla E_{Fn}(\mathbf{r}) - \frac{qD_n}{k_BT}n(\mathbf{r})\nabla E_c(\mathbf{r}) + \mu_n n(\mathbf{r})\nabla E_c(\mathbf{r}) \qquad (3.14)
\end{aligned}$$

If we use Einstein relation (3.11), last two terms cancel out and finally obtain [3]

$$\mathbf{J}_n(\mathbf{r}) = \mu_n n(\mathbf{r})\left(\nabla E_{Fn}(\mathbf{r})\right) \qquad (3.15)$$

Similar expression can be obtained for holes and it is

$$\mathbf{J}_p(\mathbf{r}) = \mu_p p(\mathbf{r})\nabla(E_{Fp}(\mathbf{r})) \qquad (3.16)$$

The above equations for currents are commonly used to model carrier transport problem in semiconductor devices and are known as semiconductor equations. The model based on these equations is basically macroscopic theory with one of Maxwell's equation to describe the static electric field in semiconductor medium and two continuity equations to describe the movement of charged carriers using the macroscopic theory of carrier drift and diffusion. In addition, some solid-state physics also can be built into the model through terms like carrier densities p, n and recombination rate R. As indicated in the above equations, static potential ψ has entered the flux expression and carrier densities p, n appeared in the Poisson equation, thus the problem is formulated in a self-consistent way.

Under nonequilibrium conditions, for degenerate semiconductors one can obtain generalized Einstein relations [5, 6] as follows. We outline derivation for electrons. Subsitute expression (3.1) into formula for current and have

$$\mathbf{J}_n = qD_n\nabla\left[N_c F_{1/2}(\sigma_n(\mathbf{r}))\right] - q\mu_n n\nabla\psi(\mathbf{r}) \qquad (3.17)$$

Evaluate derivative in the above equation

$$\nabla \left[N_c F_{1/2} \left(\sigma_n(\mathbf{r}) \right) \right] = N_c \left[\nabla \left(\sigma_n \right) \right] \frac{\partial F_{1/2} \left(\sigma_n \right)}{\partial \sigma_n} \tag{3.18}$$

From definition (3.3), one finds

$$\left[\nabla \left(\sigma_n \right) \right] = \frac{1}{k_B T} \left[\nabla E_{Fn}(\mathbf{r}) - \nabla E_c(\mathbf{r}) \right] \tag{3.19}$$

Using general property of Fermi-Dirac integrals [7, 8]

$$\frac{\partial F_j \left(\eta \right)}{\partial \eta} = F_{j-1} \left(\eta \right)$$

one finds for argument $j = 1/2$

$$\frac{\partial F_{1/2} \left(\sigma_n \right)}{\partial \sigma_n} = F_{-1/2} \left(\sigma_n \right) \tag{3.20}$$

Substituting (3.18), (3.19) and (3.20) into current equations (3.17), and using Eq. (3.12) yields

$$
\begin{aligned}
\mathbf{J}_n &= qD_n \frac{1}{k_B T} \left[\nabla E_{Fn}(\mathbf{r}) - \nabla E_c(\mathbf{r}) \right] F_{-1/2} \left(\sigma_n(\mathbf{r}) + \mu_n n \nabla E_c(\mathbf{r}) \right. \\
&= \frac{qD_n}{k_B T} n \frac{F_{-1/2} \left(\sigma_n(\mathbf{r}) \right)}{F_{1/2} \left(\sigma_n(\mathbf{r}) \right)} \nabla E_{Fn}(\mathbf{r}) - \\
&\quad -n \nabla E_c(\mathbf{r}) \left\{ \frac{qD_n}{k_B T} \frac{F_{-1/2} \left(\sigma_n(\mathbf{r}) \right)}{F_{1/2} \left(\sigma_n(\mathbf{r}) \right)} - \mu_n \right\}
\end{aligned} \tag{3.21}
$$

If we assume that the last term in the curly brackets is zero, one obtains the relation between mobility and diffusion coefficient

$$\mu_n = \frac{qD_n}{k_B T} \frac{F_{-1/2} \left(\sigma_n \right)}{F_{1/2} \left(\sigma_n \right)}$$

or

$$D_n = \frac{k_B T}{q} \mu_n \frac{F_{1/2} \left(\sigma_n \right)}{F_{-1/2} \left(\sigma_n \right)} \tag{3.22}$$

which is consistent with (3.15). For $n < N_c$ this equation may be approximated as [3]

$$
\begin{aligned}
D_n &= \frac{\mu_n k_B T}{q} 1 + 0.35355 \left(\frac{n}{N_c} \right) - 9.9 \times 10^{-3} \left(\frac{n}{N_c} \right)^2 + \\
&\quad + 4.45 \times 10^{-4} \left(\frac{n}{N_c} \right)^3 + \dots
\end{aligned} \tag{3.23}
$$

The expression (3.22) is known as the generalized Einstein relation and it is a generatization of expression (3.11) to the degenerate situations [3, 6]. Using relation (3.22) in (3.21), one finally obtains expression for electron's current in terms of quasi-Fermi energy [3, 5] which is the same as (3.15).

$$\mathbf{J}_n(\mathbf{r}) = \mu_n n \nabla(E_{Fn}(\mathbf{r})) \tag{3.24}$$

Similarly, for holes

$$\mathbf{J}_p(\mathbf{r}) = \mu_p p \nabla(E_{Fp}(\mathbf{r})) \tag{3.25}$$

Here modified Einstein relation for holes has been used

$$D_p = \frac{k_B T}{q} \mu_p \frac{F_{1/2}(\sigma_p)}{F_{-1/2}(\sigma_p)} \tag{3.26}$$

The expressions for current fluxes in (3.24), (3.25) are given in terms of the gradients of quasi-Fermi levels, implying the continuity of the quasi-Fermi levels, therefore they only apply for the regions of slowly varying materials composition. The current fluxes across hetero-interfaces are formalized in a different way and are described below.

3.2   Concentrations in Non-Equilibrium Situations

In this section, we will express electron and hole concentrations in terms of electrostatic potential and also heterostructure potentials. Such relations are useful when choosing set of independent variables to solve electrical problem and also in solving Poisson equation.

We remaind here the Poisson equation which links electrostatic potential and densities of carriers. It is written as

$$\nabla \cdot (\varepsilon_r(\mathbf{r}) \nabla \psi) = -\frac{q}{\varepsilon_0}(p - n + C_{dop}) \tag{3.27}$$

where $C_{dop} = N_D^+ - N_A^-$ describes net doping and ψ is the electrostatic potential. To solve for potential ψ one therefore needs expressions for ψ in terms of densities of electrons and holes, p and n. We start with non-degenerate situation described by Boltzmann statistics.

3.2.1   Boltzmann statistics

In non-equilibrium, we generalize previous relations (2.28) and (2.29) by introducing quasi-Fermi energies E_{Fn} and E_{Fp} and express concentrations of electrons and holes as

$$n = n_{in} \exp\left[\frac{E_{Fn} - E_i}{k_B T}\right] \tag{3.28}$$

and

$$p = n_{in} \exp\left[\frac{E_i - E_{Fp}}{k_B T}\right] \tag{3.29}$$

Expression for E_i has been derived previously and it is (see Eq.(2.50))

$$\begin{aligned}E_i(x) &= E_0 - \chi_e(x) - q\psi(x) - \frac{1}{2}E_g(x) + \frac{k_B T}{2}\ln\frac{N_v}{N_c} \\ &= const - q\psi(x)\end{aligned} \tag{3.30}$$

In the last step we assumed homostructure where electron affinity and bandgap are constant. Introduce quasi-Fermi potentials ϕ_n and ϕ_p for electrons and holes, respectively, as

$$E_{Fp} = q\phi_p \tag{3.31}$$

$$E_{Fn} = q\phi_n \tag{3.32}$$

Substitute the above definitions of quasi-Fermi potentials into expressions (3.29) and (3.28) and use Eq.(3.30) and obtains [9, 10]

$$\begin{aligned}n &= n_i \exp\left[\frac{q}{k_B T}\left(\phi_n - const + \psi\right)\right] \\ &= n_i \exp\left[\frac{q}{k_B T}\left(\psi - \phi_n'\right)\right]\end{aligned} \tag{3.33}$$

The above expression is true for homostructure in non-equilibrium in the non-degenerate limit. Similarly, for holes one finds

$$p = n_i \exp\left[\frac{q}{k_B T}\left(\phi_p' - \psi\right)\right] \tag{3.34}$$

In the above equations, we have introduced new (shifted) quasi-Fermi potentials

$$\phi_p' = const - \phi_p$$

and

$$\phi_n' = const - \phi_n$$

3.2.2   Fermi-Dirac statistics

Now we outline derivations of electron and hole concentrations under non-equilibrium conditions assuming degenerate limit. We provide detailed derivation for electrons. For holes one takes similar steps. In degenerate case when deriving expression for electron concentration one has to use Fermi-Dirac statistics which results in appearance of the Fermi-Dirac integral $F_{1/2}(\eta)$. Such functions are more difficult for numerical implementation. To simplify numerical implementation, one therefore aims to write concentrations in the Boltzmann-like form which is better suited for standard numerical methods.

We start, following Lundstrom and Schuelke [11] by introducing quasi-Fermi potentials as (note the difference with the previous definitions)

$$\phi_n = \frac{E_F - E_{Fn}}{q} \tag{3.35}$$

$$\phi_p = \frac{E_F - E_{Fp}}{q} \tag{3.36}$$

where E_F is the (equilibrium) Fermi energy and E_{Fn} and E_{Fp} are the quasi-Fermi levels for electrons and holes. Using the above definition of ϕ_n, from (2.44) and (3.3), one finds

$$
\begin{aligned}
\sigma_n &= \frac{1}{k_B T} \left\{ E_F - q\phi_n - q\psi_0 + q\chi_e + q\psi \right\} \\
&= \frac{q}{k_B T} \left\{ \psi - \phi_n + \chi_e \right\}
\end{aligned} \tag{3.37}
$$

where we introduced potential ψ_0 corresponding to a vacuum as $E_0 = q\psi_0$ and choose $q\psi_0 = E_F$.

We also assume that Boltzmann statistics could be applied for intrinsic semiconductor [12] which results in the following relation, see Eq.(2.31)

$$n_i = \sqrt{N_c N_v} \exp\left(-\frac{E_g}{2k_B T} \right) \tag{3.38}$$

The derivation starts with the expression for electron density in degenerate case, Eq. (3.1). One writes it as (dropping position's dependence)

$$1 = \frac{N_c}{n} F_{1/2}(\sigma_n)$$

Taking logarithm of the above and performing indicated algebraic steps gives

$$
\begin{aligned}
0 &= \ln 1 = \ln\left[\frac{N_c}{n} F_{1/2}(\sigma_n) \right] \\
&= \ln\left[\frac{n_i}{n} \frac{N_c}{n_i} F_{1/2}(\sigma_n) \right] \\
&= -\ln\frac{n}{n_i} + \ln F_{1/2}(\sigma_n) + \ln\frac{N_c}{n_i}
\end{aligned} \tag{3.39}
$$

The last term can be arranged as follows

$$
\begin{aligned}
\ln\frac{N_c}{n_i} &= -\ln\frac{n_i}{\sqrt{N_c N_v}} \sqrt{\frac{N_v}{N_c}} \\
&= -\ln\frac{n_i}{\sqrt{N_c N_v}} - \ln\sqrt{\frac{N_v}{N_c}} \\
&= \frac{E_g}{2k_B T} - \ln\sqrt{\frac{N_v}{N_c}}
\end{aligned}
$$

where we have used relation (3.38). Using the last result, Eq. (3.39) takes the form

$$
\begin{aligned}
\ln \frac{n}{n_i} &= \frac{E_g}{2k_BT} - \frac{1}{2}\ln\frac{N_v}{N_c} + \ln F_{1/2}(\sigma_n) \\
&= \frac{q}{k_BT}\left[\frac{E_g}{2q} - \frac{k_BT}{2q}\ln\frac{N_v}{N_c} + \frac{k_BT}{q}\ln F_{1/2}(\sigma_n)\right]
\end{aligned}
$$

Now, we add and subtract terms as follows

$$
\begin{aligned}
\ln \frac{n}{n_i} &= \frac{q}{k_BT}\left[\psi - \phi_n + \frac{\chi_e}{q} + \frac{E_g}{2q} - \frac{k_BT}{2q}\ln\frac{N_v}{N_c}\right. \\
&\qquad \left. + \frac{k_BT}{q}\ln F_{1/2}(\sigma_n) - \psi + \phi_n - \frac{\chi_e}{q}\right] \\
&= \frac{q}{k_BT}\left(\psi - \phi_n + \frac{\theta}{q} + \frac{\gamma_n}{q}\right)
\end{aligned}
$$

From above, we finally have [11, 13]

$$
n = n_i \exp\left[\frac{q}{k_BT}\left(\psi - \phi_n + \frac{\theta}{q} + \frac{\gamma_n}{q}\right)\right] \tag{3.40}
$$

In the above we have introduced

$$
\theta = \chi_e + \frac{E_g}{2} - \frac{k_BT}{2}\ln\frac{N_v}{N_c} \tag{3.41}
$$

and also

$$
\begin{aligned}
\gamma_n &= k_BT\ln F_{1/2}(\sigma_n) - q\left(\psi - \phi_n + \frac{\chi_e}{q}\right) \\
&= k_BT\left\{\ln F_{1/2}(\sigma_n) - \frac{q}{k_BT}\left[\psi - \phi_n + \frac{\chi_e}{q}\right]\right\} \\
&= k_BT\left[\ln F_{1/2}(\sigma_n) - \sigma_n\right] \\
&= k_BT\ln\frac{\ln F_{1/2}(\sigma_n)}{\exp(\sigma_n)} \tag{3.42}
\end{aligned}
$$

where we applied elementary relation $x = e^{\ln x}$ to the last term and used definition (3.37).

This notation preserves the Boltzmann-like formulation for electron concentration which is better suited for standard numerical methods while using Fermi-Dirac statistics. Similar steps can be taken for holes. The resulting expression is

$$
p = n_i \exp\left\{\frac{q}{k_BT}\left(\phi_p - \psi - \frac{\theta}{q} + \frac{\gamma_p}{q}\right)\right\} \tag{3.43}
$$

with

$$
\gamma_p \equiv k_BT\ln\frac{F_{1/2}(\sigma_p)}{\exp(\sigma_p)} \tag{3.44}
$$

Here, σ_p is defined by (3.4).

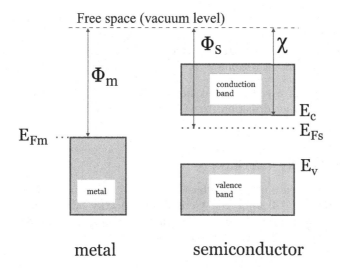

FIGURE 3.1
Metal and semiconductor before contact.

3.3 Contacts

In order to solve device equations, proper boundary conditions must be formulated. The boundary conditions for electrical equations include ohmic contacts, Schottky contacts, Neumann (reflective) boundaries and current controlled contacts. For the wave equation, the boundary conditions are zero boundary, reflective boundary with even and odd symmetries and exponential boundary.

In this section, we will concentrate on electrical boundary conditions and will provide information on Schottky barrier, ohmic contacts and currents across heterointerface. Contacts are extensively discussed in literature. Important references include: [3, 6, 14–16], to just name a few.

Metal and semiconductor before contact are shown in Fig. 3.1. Fermi energies, work functions in a metal and semiconductor and band-edge energies are shown.

Basic terminology associated with metal-semiconductor (MS) contacts is [6]:

- Metal-semiconductor contacts can be rectifying or nonrectifying depending on the properties of both materials and the preparation method of making contact.

- Rectifying is a contact which has a larger barrier to charge carrier transport in one direction than in the opposite direction. Such metal-semiconductor contacts display asymmetric current-voltage characteristics, i.e. allowing high current to flow across under the forward bias condition and blocking current off under the reverse bias; this behavior is controlled by the bias voltage dependent changes of the potential barrier height in the contact region

• Nonrectifying is a contact that has no effective barrier to charge carrier transport in either direction. Usually contacts with no barrier are referred to as ohmic.

3.3.1   Schottky barriers

Schottky barriers are important basic concept and has been discussed widely in literature, see [14] and [16]. Generally, Schottky effect refers to barrier lowering in a metal-semiconductor contact [15]. The resulting devices which use such contacts are called Schottky diodes [6].

Work function $q\Phi_m$ of a metal in a vacuum is the energy required to remove an electron from the Fermi level to a vacuum outside the metal. When one brings negative charges near the metal surface, positive (image) charges are induced in the metal. When the created image force is combined with an applied electric field, the effective work function is reduced. Such barrier lowering is called the Schottky effect [15]. The terminology is applied to all metal-semiconductor contacts. Formation of Schottky barrier is shown in Fig. 3.2 for metal-n type semiconductor and metal-p type semiconductor with

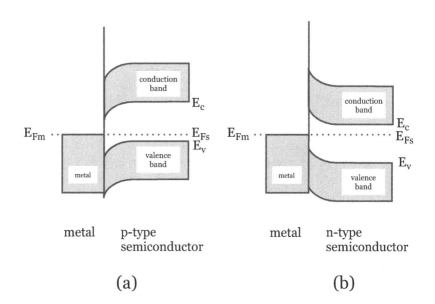

FIGURE 3.2
Schottky contacts.

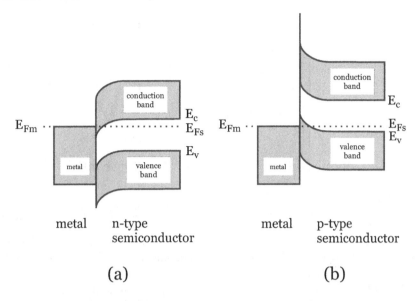

metal   n-type
         semiconductor

metal   p-type
         semiconductor

(a)   (b)

FIGURE 3.3
Possible metal-semiconductor junction configurations which lead to ohmic contacts.

no applied bias. In equilibrium with no applied bias, there are two drift currents of equal magnitude but opposite directions, which flow across the junction because of the built-in electric field.

3.3.2 Ohmic contact

Ohmic contact between metal and semiconductor is characterized by a very low resistance independent of applied voltage (may be represented by constant resistance). To form an "ohmic" contact, metal and semiconductor must be selected such that there is no potential barrier formed at the interface (or potential barrier is so thin that charge carriers can readily tunnel through it). Formation of ohmic contacts is illustrated in Fig. 3.3.

Ohmic contacts show a linear $I - V$ characteristics in both biasing directions. Comparison of $I - V$ characteristics of ohmic and Schottky contacts is shown in Fig. 3.4 (after Shur [3]).

3.4 Currents across Heterointerface

Transport over heterointerface has been discussed by several groups [17–25] and more recently by Calado et al [26]. Both non-degenerate and degenerate

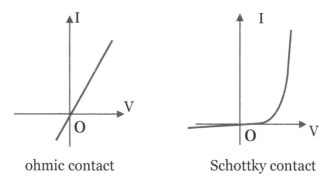

ohmic contact   Schottky contact

FIGURE 3.4
Comparison of current-voltage characteristics of ohmic and Schottky contacts.

situations were considered. Also various models of transport have been created. Among the possible conduction mechanisms over the heterojunction the most often considered were thermionic and tunnel emission. The heterointerface is modelled as abrupt junction since most physical quantities are discontinuous across an abrupt heterojunction. An abrupt heterojunction exhibits non-ohmic behavior. In the present work, we apply the thermionic emission model [27] to describe current transport across heterointerface.

3.4.1   Thermionic model

Here we discuss classical transport over heterojunction barrier and concentrate on one-dimensional case. We follow Perlman and Feucht [17] and consider two materials forming heterostructure with a potential barrier, as shown in Fig. 3.5.

We assume medium 1, on the left side with an effective mass of m_1 and medium 2 (on the right) with an effective mass m_2. Potential step is located between two points x_1 and x_2.

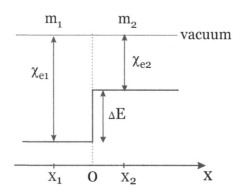

FIGURE 3.5
Definitions used in determining flux across heterointerface.

Potential ΔE is expressed in terms of electron affinities for both regions

$$\Delta E = q\left(\chi_{e1} - \chi_{e2}\right) \tag{3.45}$$

The analysis relies on a rigorous classical treatment based on the kinetic approach. In this approach, the Maxwellian distribution function $f(v_x)$ is

$$f(v_x) = \left(\frac{m}{2\pi k_B T}\right)^{1/2} \exp\left(-\frac{m}{2k_B T}v_x^2\right) \tag{3.46}$$

Define thermal velocity v_T as

$$v_T = \int_0^\infty dv_x v_x f(v_x) \tag{3.47}$$

Applying Maxwellian distribution (3.46), the thermal velocity v_T is evaluated to be

$$
\begin{aligned}
v_T &= \left(\frac{m}{2\pi k_B T}\right)^{1/2} \int_0^\infty dv_x v_x \exp\left(-\frac{m}{2k_B T}v_x^2\right) = \sqrt{\frac{k_B T}{2\pi m}} \\
&= \frac{1}{2}\sqrt{\frac{2k_B T}{\pi m}} = \frac{1}{2}\bar{v}
\end{aligned}
\tag{3.48}
$$

where we have applied integral formula (3.76) and introduced $\bar{v}$ as the mean velocity of the electron perpendicular to the interface [21].

Conservation law of energy in such 1D situation requires that

$$\frac{1}{2}m_1 v_{x1}^2 = \frac{1}{2}m_2 v_{x2}^2 + \Delta E \tag{3.49}$$

Here v_{x1} represents 1D component of velocity at point $x = x_1$ in medium 1 and v_{x2} represents 1D component of velocity at point $x = x_2$ in medium 2. Particle flux in 1D from between point x_1 and x_2 is evaluated by an integration of the total number of particles which approach the barrier with sufficient x-component of velocity to pass over barrier

$$F_{12} = n_1 \int_{v_{x1m}}^\infty dv_{x1} v_{x1} \sqrt{\frac{m_1}{2\pi k_B T}} e^{-\frac{m_1}{2k_B T}v_{x1}^2} \tag{3.50}$$

where n_1 is the density of particles at point x_1.

The minimum allowable velocity v_{x1m} is determined from Eq.(3.49) by requiring that v_{x2} be greater than or equal to some value which is determined by the barrier configuration. Particle can leave barrier region at $x = x_2$ if its speed is $v_{x2} \geq 0$. The minimum allowable velocity v_{x1m} is obtained for $v_{x2} = 0$ and it is

$$v_{x1m}^2 = \frac{2}{m_1}\Delta E \tag{3.51}$$

Evaluating integral (3.50) using Eqs.(3.78) and (3.51) gives

$$F_{12} = \frac{1}{2} n_1 \bar{v}_2 \sqrt{\frac{m_2}{m_1}} \exp\left(-\frac{\Delta E}{k_B T}\right) \tag{3.52}$$

where $\bar{v}_2$ is the mean velocity of the electron given by Eq.(3.48).

The reversed flux F_{21} is evaluated in a similar way. There is no potential barrier in that case and flux is

$$F_{21} = \frac{1}{2} n_2 \bar{v}_2 \tag{3.53}$$

where $\bar{v}_2$ is the speed at point $x = x_2$.

Total net flux $F = F_{12} - F_{21}$ is

$$F = \frac{1}{2} n_1 \bar{v}_2 \sqrt{\frac{m_2}{m_1}} \exp\left(-\frac{\Delta E}{k_B T}\right) - \frac{1}{2} n_2 \bar{v}_2 \tag{3.54}$$

Following Li et al [21], the net flux F can be further evaluated as follows. First, notice that in 1D electron concentration is expressed as

$$n = N_c \exp\left(\frac{E_{Fn} - E_c}{k_B T}\right)$$

with $N_c = \sqrt{\frac{m^* k_B T}{2\pi\hbar^2}} = const\sqrt{m^*}$. For heterostructures, see Eq.(2.44)

$$E_c = E_0 - q\chi_e - q\psi$$

with $E_0 = q\phi_0$ being the reference energy. This allows us to write electron's concentration as

$$n = const\sqrt{m^*} \exp\left[\frac{q}{k_B T}\left(\phi_{Fn} - \phi_0 + q\chi_e + \psi\right)\right] \tag{3.55}$$

Between points x_1 and x_2 there is a change in effective masses. Using formula (3.55), the total flux can be expressed as

$$
\begin{aligned}
F &= \frac{1}{2} n_1 \bar{v}_2 \sqrt{\frac{m_2}{m_1}} e^{-\frac{1}{k_B T}(\chi_{e1} - \chi_{e2})} - \frac{1}{2} n_2 \bar{v}_2 \\
&= \frac{1}{2} n_1 \bar{v}_2 \frac{N_{c2} \exp\left(\frac{1}{k_B T}\chi_{e2}\right)}{N_{c1} \exp\left(\frac{1}{k_B T}\chi_{e1}\right)} - \frac{1}{2} n_2 \bar{v}_2 \\
&= \frac{1}{2} n_1 \bar{v}_2 \frac{N_{c3} \exp\left[\frac{q}{k_B T}\left(\phi_{Fn1} - \phi_0 + \frac{\chi_{e2}}{q} + \psi\right)\right]}{N_{c1} \exp\left[\frac{q}{k_B T}\left(\chi_{e1}\phi_{Fn1} - \phi_0 + \frac{\chi_{e1}}{q} + \psi\right)\right]} - \frac{1}{2} n_2 \bar{v}_2 \\
&= \frac{1}{2} n_1 \bar{v}_2 \frac{n_2(\phi_{Fn1})}{n_1} - \frac{1}{2} n_2 \bar{v}_2 \\
&= v_{T2}\left[n_2(\phi_{Fn1}) - n_2(\phi_{Fn2})\right] \tag{3.56}
\end{aligned}
$$

where $n_2(\phi_{Fn1})$ is the electron's concentration at point x_2 evaluated with the quasi-Fermi potential at point x_1. In the final step we also used the fact that $\bar{v}_2 = 2v_{T2}$.

In a similar way one can also derive flux for holes. The equations show that carrier at one side of the heterointerface emits to the other side of the interface with a thermal velocity.

Identical results can be obtained assuming degenerate statistics, see Li et al [21]. They derived unified expressions for current transport across heterointerface assuming Boltzmann and also Fermi statistics. Introducing parameter γ they can be expressed as [28]

$$J_n = -\gamma v_{R,A}(n_A - n_{A,0}) \tag{3.57}$$

where $n_{A,0}$ is the electron concentration on side A assuming that the Fermi energy on side A is the same as that on side B. γ_{hn} is a scaling factor to account for effects such as tunnelling. These equations ensure that, when the quasi-Fermi levels on both sides of the barrier are the same, the net current is zero. More details on the derivation of the boundary conditions of the above equations can be found in [17, 29] and [30].

3.4.2 Gradient model

The above abrupt junction model has its limitations, especially when both sides of the junction are heavily doped with the same type of dopant (e.g. donors). In such a case the Fermi level is is strongly pinned at the band edges on both sides and the potential barrier is forced to form a sharp peak above the Fermi level. Since the peak region in the barrier is strongly depleted of carriers, the resistance is very high. If this junction happens to be reverse biased (i.e. the carriers on barrier side tends to be more depleted when the bias is applied), the high resistance can cause an unrealistic large voltage drop there.

The above limitation arises due to application of drift-diffusion model as the transport mechanism. In reality, the main mechanism in the above situation will be by tunnelling and such cases need another approaches [31].

As the intermediate approach we assume a smooth transition over the hetero-region, see Calado et al [26]. We illustrate this approach in one-dimensional situation. In this model expressions for currents of electrons and holes acquire additional gradient terms due to spatial variation of $N_c, N_v, \chi_e, \Phi_{IP}$.

Energy levels relevant to this analysis are illustrated in Fig. 3.6. The following quantities are shown: E_{vac} vacuum level, χ_e electron affinity, Φ_{IP} ionization potential, E_{CB} and E_{VB} conduction and valence band energies, E_g band gap.

Since various conventions exist in the literature, for the purpose of discussing current across heterointerface we provide summary starting from basic assumptions. We follow Shur [3]. The expressions for current densities of electrons and holes within the bulk material in one-dimension are

$$J_n(x) = q \left[-n v_n(E) + D_n \frac{dn(x)}{dx} \right] \tag{3.58}$$

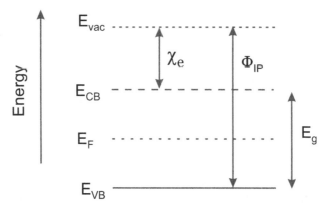

FIGURE 3.6
Semiconductor energy levels. Illustration for gradient model [26].

$$J_p(x) = q \left[p v_p(E) - D_p \frac{dp(x)}{dx} \right] \tag{3.59}$$

where E is the electric field

$$E = -\frac{d\psi(x)}{dx} \tag{3.60}$$

Drift velocities are

$$v_n(E) = -\mu_n E \tag{3.61}$$

$$v_p(E) = \mu_p E \tag{3.62}$$

Densities of carriers are expressed in terms of quasi-Fermi energies E_{Fn} and E_{Fp} as

$$n = N_c \exp \left(\frac{E_{Fn} - E_c}{k_B T} \right) \tag{3.63}$$

$$p = N_v \exp \left(\frac{E_v - E_{Fp}}{k_B T} \right) \tag{3.64}$$

with the expressions of effective densities of states

$$N_c = 2 \left(\frac{m_n^* k_B T}{2\pi \hbar^2} \right)^{3/2} \tag{3.65}$$

and

$$N_v = 2 \left(\frac{m_p^* k_B T}{2\pi \hbar^2} \right)^{3/2} \tag{3.66}$$

The above are position-dependent due to position dependence of effective masses.

From above formulas one obtains expressions for quasi-Fermi energies

$$E_{Fn} = E_c + k_B T \ln \frac{n}{N_c} \qquad (3.67)$$

$$E_{Fp} = E_v - k_B T \ln \frac{p}{N_v} \qquad (3.68)$$

Using the above formulas and Einstein relations $D = \mu k_B T/q$, we can obtain electrical currents. Additionally, we also use the following relations, as shown in Fig. 3.6

$$E_{vac} = -q\psi, \quad E_c = E_{vac} - \chi_e \implies E_c = -q\psi - \chi_e \qquad (3.69)$$

and

$$E_{vac} = -q\psi, \quad E_v = E_{vac} - \phi_{IP} \implies E_v = -q\psi - \phi_{IP} \qquad (3.70)$$

Current densities for electrons and holes are determined as

$$J_n = \mu_n n \frac{\partial E_{Fn}}{\partial x} \qquad (3.71)$$

and

$$J_p = \mu_p p \frac{\partial E_{Fp}}{\partial x} \qquad (3.72)$$

As indicate above, due to space dependence of $\Phi_{EA}, \Phi_{IP}, N_{CB}$ and N_{VB} for heterostructures, the modified equations for heterostructures become [26, 32]

$$J_n = \mu_n n \left(qE - \frac{\partial \Phi_{EA}}{\partial x} - \frac{k_B T}{N_c} \frac{\partial N_c}{\partial x} \right) + k_B T \mu_n \frac{\partial n}{\partial x} \qquad (3.73)$$

$$J_p = \mu_p p \left(qE - \frac{\partial \Phi_{IP}}{\partial x} + \frac{k_B T}{N_v} \frac{\partial N_v}{\partial x} \right) - k_B T \mu_p \frac{\partial p}{\partial x} \qquad (3.74)$$

In order to determine gradients of the above quantities a finite thickness of the interface must be chosen. Typical values are between 1 and 2 nm [26].

3.5 Appendix

Here we summarize some useful results of integration used in this chapter. Those are

$$I_0(a) = \frac{1}{2} \sqrt{\frac{\pi}{a}} \qquad (3.75)$$

$$I_1(a) = \frac{1}{2a} \qquad (3.76)$$

where we have used standard definition

$$I_n(a) = \int_0^\infty dx \, x^n \, e^{-ax^2} \qquad (3.77)$$

In evaluating fluxes, we will also need the integral (evaluated using Maple 9.5)

$$\int_b^\infty dx \; x \; e^{-ax^2} = \frac{1}{2a} e^{-ab^2} \tag{3.78}$$

Bibliography

[1] M. Lundstrom. *Fundamentals of Carrier Transport. Second Edition.* Cambridge Unoversity Press, Cambridge, UK, 2000.

[2] A.I. Fedoseyev, M. Turowski, and M.S. Wartak. Kinetic and quantum models for nanoelectronic and optoelectronic device simulation. *J. Nanoelectron. Optoelectron.*, 2:234–256, 2007.

[3] M. Shur. *Physics of Semiconductor Devices.* Prentice Hall, Englewood Cliffs, NJ, 1990.

[4] G.H. Song. *Two-dimensional simulation of quantum-well lasers including energy transport.* PhD thesis, University of Illinois at Urbana-Champaign, 1990.

[5] R.A. Smith. *Semiconductors. Second Edition.* Cambridge University Press, Cambridge, 1978.

[6] C.M. Wolfe, N. Holonyak, Jr., and G.E. Stillman. *Physical Properties of Semiconductors.* Prentice Hall, Englewood Cliff, New Jersey, 1989.

[7] J. McDougall and E.C. Stoner. The computation of Fermi-Dirac functions. *Phil. Trans. Roy. Soc. London*, 237:67–104, 1938.

[8] J.S. Blakemore. Approximations for Fermi-Dirac integrals, especially the function $F_{1/2}(\eta)$ used to describe electron density in a semiconductor. *Solid State Electron.*, 25:1067–1076, 1982.

[9] M. Kurata. *Numerical Analysis for Semiconductor Devices.* Lexington-Books, New York, 1982.

[10] W. Fichtner, D.J. Rose, and R.E. Bank. Semiconductor device simulation. *IEEE Trans. Electron Devices*, 30:1018–1030, 1983.

[11] M.S. Lundstrom and R.J. Schuelke. Numerical analysis of heterostructure semiconductor devices. *IEEE Trans. Electron Devices*, ED-30:1151–1159, 1983.

[12] M.S. Lundstrom and R.J. Schuelke. Modeling semiconductor heterojunctions in equilibrium. *Solid State Electron.*, 25:683–691, 1982.

[13] A. Champagne, R. Maciejko, and J.M. Glinski. The performance of double active region InGaAsP lasers. *IEEE J. Quantum Electron.*, 27:2238–2247, 1991.

[14] R.A. Pierret. *Semiconductor Device Fundamentals.* Addison-Wesley, Reading, Massachusetts, 1996.

[15] B.G. Streetman and S. Banerjee. *Solid State Electronic Devices.* Prentice-Hall, Upper Saddle River, New Jersey, 2000.

[16] K. Hess. *Advanced theory of semiconductor devices.* IEEE Press, New York, 2000.

[17] S.S. Perlman and D.L. Feucht. p-n heterojunctions. *Solid State Electron.*, 7:911–923, 1964.

[18] C.R. Crowell and Sze S.M. Current transport in metal-semiconductor barriers. *Solid State Electron.*, 9:1035–1048, 1966.

[19] S. Motted and J.E. Viallet. Thermionic emission in heterojunctions. In G. Baccarani and M. Rudan, editors, *Simulation of semiconductor devices and processes*, volume 3, pages 97–108. Tecnoprint, Bologna, 1988.

[20] K.M. Chang. A consistent model for carrier transport in heavily doped semiconductor devices. *Semicond. Sci. Technol.*, 3:766–772, 1988.

[21] Z.-M. Li, S.P. McAlister, and C.M. Hurd. Use of Fermi statistics in two-dimensional numerical simulation of heterojunction devices. *Semicond. Sci. Technol.*, 5:408–413, 1990.

[22] M. Grupen, K. Hess, and G.H. Song. Simulation of transport over heterojunctions. In W. Fichtner and D. Aemmer, editors, *Simulation of semiconductor devices and processes*, volume 4, pages 303–311. Hartung-Gorre Verlag, Konstanz, 1991.

[23] K.-M. Chang, J.-Y. Tsai, and C.-Y. Chang. New physical formulation of the thermionic emission current at the heterojunction interface. *IEEE Electron Device Letters*, 14:338–341, 1993.

[24] M.A. Stettler and M.S. Lundstrom. A detailed investigation of heterojunction transport using a rigorous solution to the Boltzmann equation. *IEEE Trans. Electron Devices*, 41:592–600, 1994.

[25] M. Horak. Conservation laws and charge transport across heterojunction barriers. *Solid State Electron.*, 42:269–276, 1998.

[26] P. Calado, I. Gelmetti, B. Hilton, M. Azzouzi, J. Nelson, and P.R.F. Barnes. Driftfusion: an open source code for simulating ordered semiconductor devices with mixed ionic-electronic conducting materials in one dimension. *Journal of Computational Electronics*, 21:960–991, 2022.

[27] S.M. Sze. *Physics of Semiconductor Devices*. Wiley, New York, 1969.

[28] Z.-M. Li. Two-dimensional numerical simulation of semiconductor lasers. In W.P. Huang, editor, *Electromagnetic Waves, Methods for Modeling and Simulation of Guided-Wave Optoelectronic Devices: Part II:Waves and Interactions*, volume PIER 11, pages 301–344. 1995.

[29] R.J. Schuelke and M.S. Lundstrom. Thermionic emission-diffusion theory of isotype heterojunctions. *Solid State Electron.*, 27:1111–1116, 1984.

[30] A. Champagne. *Modelisation des lasers InGaAsP-InP a double heterostructure et a double region active*. PhD thesis, Departement de Genie Physique, Ecole Polytechnique, Universite de Montreal, 1992.

[31] A.I. Fedoseyev, A. Przekwas, M. Turowski, and M.S. Wartak. Robust computational models of quantum transport in electronic devices. *Journal of Computational Electronics*, 3:231–234, 2004.

[32] Philip Calado. *Transient Optoelectronic. Characterisation and Simulation of Perovskite Solar Cells*. PhD thesis, Imperial College London, 2018.

4

p-n Junctions

The p-n junction is the fundamental building block in electronics and opto-electronics. It found applications in many diodes, like: light emitting diodes (LED), semiconductor lasers, photodiodes, solar cells, Zener diodes. Here we provide a condensed summary of operation of homo- and hetero-junctions. Our description starts with the simplest model, known as "depletion approximation". Required information about densities of electrons and hole were summarized in the previous chapter.

In this chapter we concentrate on the following:

- homogeneous p-n junction

- homogeneous p-i-n junction

- hetero junctions

First, lets try to answer the question of "what is p-n junction?" Possible answer relates to how p-n junctions are formed. The simplest, the so-called homojunction is formed between semiconductors that differ in their doping levels but not in their atomic or alloy compositions. Both regions have the same bandgap energy.

To be specific, consider silicon (Si) cylinder where part of it is doped, say by boron (B) which makes it p-type (N_A is the acceptor concentration). When we doped the adjacent region with, say phosphorus (P) it will become n-type (N_D is the donor concentration).

In p-type region $N_A > N_D$ and in n-type region $N_A < N_D$. The point where $N_A = N_D$ is known as metallurgical junction [1].

To model it, one usually assumes equilibrium conditions and also that the p-n junction is abrupt. Then, in the absence of any applied bias there is no current in the system and the Fermi level is uniform throughout the structure.

4.1 Formation of p-n Homojunction

In Fig. 4.1 we showed arrangement of bands of two semiconductors before forming pn-junction and after junction was formed. In Fig. 4.2 we show schematically another picture of p–n junction in thermal equilibrium with

DOI: 10.1201/9781003265849-4

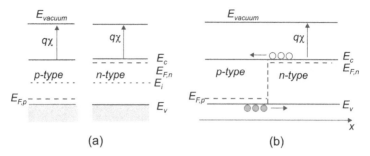

FIGURE 4.1
Energy band diagram of a p-n junction (a) before and (b) after merging the n-type and p-type regions.

zero-bias voltage applied. We indicated electron and hole concentrations, region of charge neutrality and regions which are positively and negatively charged. Due to charge redistribution internal electric field is created which is shown at the bottom. There exist electrostatic forces on electrons and holes and the direction in which the diffusion tends to move electrons and holes.

Before semiconductor materials are 'joined', the n region has a large concentration of electrons and a few holes. The reverse is true for p region. The situation after merging is shown in Fig. 4.3.

The diffusion of carriers took place as there existed a large difference of concentration of carriers. Specifically, holes diffused from the p-side into the n-side and electrons diffused from n-side to p-side. As a result of diffusion of charges a large electric field was created at the junction which stopped the diffusion process. In this way an equilibrium had been established with no further flow of current.

In the following we conduct detailed analysis of the equilibrium conditions for hole current. For electrons, one can perform the similar steps.

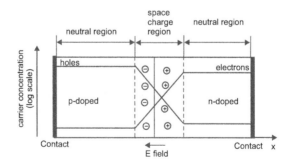

FIGURE 4.2
pn-junction- in thermal equilibrium

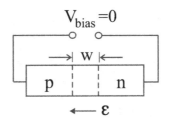

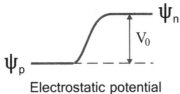

Electrostatic potential

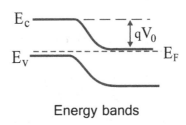

Energy bands

FIGURE 4.3
Illustration of contact potential V_0 for a pn junction in equilibrium.

Hole current consists of two components, drift and diffusion. In equilibrium they cancel each other, as

$$J_p(x) = q\left[\mu_p p(x)E(x) - D_p \frac{dp(x)}{dx}\right] = 0 \qquad (4.1)$$

Use Einstein relation and replace electrostatic field by a potential and obtain

$$-\frac{q}{k_B T}\frac{d\psi}{dx} = \frac{1}{p(x)}\frac{dp(x)}{dx} \qquad (4.2)$$

Assume that potentials and hole concentrations at the the edges of the junctions are ψ_p, ψ_n and p_p, p_n. Integration of the above equation gives

$$-\frac{q}{k_B T}\int_{\psi_p}^{\psi_n} d\psi = \int_{p_p}^{p_n}\frac{dp}{p(x)}$$

or

$$-\frac{q}{k_B T}(\psi_n - \psi_p) = \ln p_n - \ln p_p = \ln\frac{p_n}{p_p} \qquad (4.3)$$

One defines contact potential (or build-in potential) V_0 as, as shown in Fig. 4.3.

$$V_0 = \psi_n - \psi_p = \frac{k_B T}{q} \ln \frac{p_p}{p_n} \tag{4.4}$$

The build-in potential V_{bi} responsible for this electric field is

$$V_0 = V_{bi} = E_{Fn0} - E_{Fp0} \tag{4.5}$$

where E_{Fn0} and E_{Fp0} are Fermi energies of n and p regions before making contact.

4.2 Simple Model of Homojunction: Debye Length

4.2.1 n-region

Lets start with the simple discussion. We concentrate first on n-region only. The describe junction using Poisson equation. In terms of $E_c(x)$ Poisson equation becomes

$$\frac{d^2 E_c(x)}{dx^2} = \frac{q\rho(x)}{\varepsilon} \tag{4.6}$$

Potential $\psi(x)$ is related to the conduction band energy $E_c(x)$ as (see Eq.(2.44))

$$q\psi(x) = -E_c(x) + const \tag{4.7}$$

Charge density at point x is

$$\rho(x) = q\left[N_D - n(x)\right] \tag{4.8}$$

where $n(x)$ is the concentration of electrons at point x. Its expression (derived previously) is

$$n(x) = N_c \exp\left[-\frac{E_c(x) - E_F}{k_B T}\right]$$

Far away from the junction (i.e. for $x \longrightarrow -\infty$), from the above one has

$$n(x \longrightarrow -\infty) = N_D = N_c e^{-\frac{E_c(x \longrightarrow -\infty) - E_F}{k_B T}}$$

Using the above one has

$$
\begin{aligned}
n(x) &= N_c e^{-\frac{E_c(x) - E_c(x \longrightarrow -\infty)}{k_B T}} e^{-\frac{E_c(x \longrightarrow -\infty) - E_F}{k_B T}} \\
&= N_D e^{-\frac{E_c(x) - E_c(x \longrightarrow -\infty)}{k_B T}} \\
&= N_D e^{-\varphi_n(x)}
\end{aligned}
$$

where we have introduced the following parameter

$$\varphi_n(x) = \frac{E_c(x) - E_c(x \longrightarrow -\infty)}{k_B T} \tag{4.9}$$

Finally, charge density takes the form

$$\rho(x) = N_D \left[1 - e^{-\varphi_n(x)} \right] \tag{4.10}$$

The above relation indicates strong position dependence of charge density. Using relations (4.9) and (4.10) we can write Poisson equation for parameter $\varphi_n(x)$ in the case of a uniformly doped n-type

$$\frac{d^2\varphi_n(x)}{dx^2} = \frac{q}{\varepsilon k_B T} N_D \left[1 - e^{-\varphi_n(x)} \right]$$

Assuming small bending, i.e. $\varphi_n(x) \ll 1$, one can expand

$$e^{-\varphi_n(x)} \simeq 1 - \varphi_n(x)$$

Equation takes the form

$$\frac{d^2\varphi_n(x)}{dx^2} = \frac{q}{\varepsilon k_B T} N_D \varphi_n(x)$$

which has the solution

$$\varphi_n(x) = \exp\left(\pm \frac{x}{L_D} \right)$$

where

$$L_D = \sqrt{\frac{\varepsilon k_B T}{q^2 N_D}}$$

is known as the Debye length.

4.2.2 p-region

One can repeat the above discussion for a p-type region assuming uniform doping. The parameter $\varphi_p(x)$ which describes band bending in p-region is

$$\varphi_p(x) = -\frac{E_v(x) - E_v(x \longrightarrow -\infty)}{k_B T} \tag{4.11}$$

The obtained density is

$$\rho(x) = -qN_A + p(x) = -qN_A \left[1 - e^{-\varphi_p(x)} \right] \tag{4.12}$$

In this region (p-type) positive φ_p corresponds to valence band bending downwards.

4.3   Homo Junction in the Depletion Approximation

Consider pn-junction in equilibrium, i.e. when two types, i.e. n and p are connected and when external potential difference between both regions is zero. At that moment the conduction and valence band edges line up, while the Fermi levels exhibit a discontinuity at the junction. Immediately after making contact electrons start to diffuse from the n-region to the p-region and similarly holes diffuse to the n-region. As a result there will be build of extra electrons in the p-region and also extra holes in the n-region. These charges will create an electric field which will eventually stop diffusions. This will happen when Fermi level becomes constant across the junction which also signals no flow of current across the junction.

Common assumption behind the depletion approximation [1] is that the impurity concentration in semiconductor changes abruptly from acceptor impurities (N_A) to donor impurities (N_D) as shown in Fig. 4.4. It can be summarized as follows:

1. The carrier concentrations are assumed to be negligible compared to the net doping concentrations in a region of metallurgical junction,

2. The charge density outside the depletion region is taken to be identically zero.

3. Deep in the n-type region to the right and in the p-type region to the left the semiconductor remains almost neutral.

4. There is a central region which is "depleted" of carriers. Electrons have left the region $0 \leq x \leq x_n$, holes have left the region $-x_p \leq x < 0$. This central region is called the "depletion region" of the junction.

We consider one-dimensional model. Detailed description is based on Poisson equation. Our goal is to relate band bending to the electric field. To describe junction we start at the n-type side and assume:

1. uniform doping, i.e. donor concentration N_D is position independent

2. all donors are fully ionized

3. semiconductor is non-degenerate, i.e. $N_D << N_C$.

Applying the above model assumptions and solving Poisson's equation one can determine all details of the junctions.

4.3.1   Mathematical Details of the 1D Model

The physical picture of an abrupt p-n junction in thermal equilibrium is summarized in Fig. 4.4 (see [1] and [2] for more discussion). The results for 1D model are shown. Those include: doping profile, electric field, electrostatic

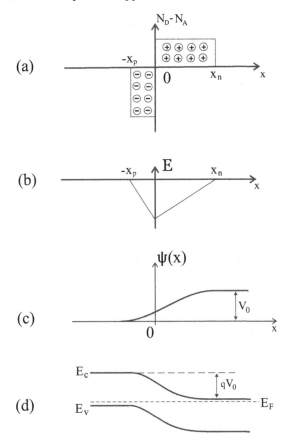

FIGURE 4.4
Abrupt p-n junction in thermal equilibrium in the depletion approximation. a) Step junction profile. b) Electric field as a function of position. c) Electrostatic potential as a function of position. d) Equilibrium energy band diagram.

potential and equilibrium energy band diagram are shown as a function of position. Those quantities are determined by solving Poisson equation with appropriate boundary conditions. V_{bi} is the total potential drop across the junction. The junction forms the basis for future generalizations, especially for heterostructures.

Charge density around metallurgical junction consists of contributions from free electrons, holes and also doping and is

$$\rho(x) = q\left(p - n + N_D^+ - N_A^-\right) \tag{4.13}$$

where q is absolute value of elementary charge, p, n are densities of holes and electrons.

We assume complete ionization of impurities, i.e.

$$N_D^+ = N_D, \qquad N_A^- = N_A \tag{4.14}$$

where N_D, N_A are doping densities. In the space charge region free carriers p and n are depleted and charge density is approximated as

$$\rho(x) = \begin{cases} -qN_A & -x_p < x < 0 \\ qN_D & 0 < x < x_n \end{cases} \tag{4.15}$$

We assumed charge distribution due to doping as shown in Fig. 4.4. It is therefore assumed that no free charges are present within the depletion region. Charges are (Fig. 4.4):

$$Q_n = qN_D x_n, \qquad Q_p = -qN_A x_p \tag{4.16}$$

4.3.1.1   Solution for the electric field

To determine electric field we use Poisson equation

$$\frac{dE(x)}{dx} = -\frac{\rho(x)}{\varepsilon} = \begin{cases} -\frac{qN_A}{\varepsilon_p \varepsilon_0} & -x_p < x < 0 \\ \frac{qN_D}{\varepsilon_n \varepsilon_0} & 0 < x < x_n \end{cases} \tag{4.17}$$

In the above we assumed that doping affects dielectric constant and their values are: ε_p in p-region and ε_n in n-region. Integrating the above one finds for the p-region

$$E(x) = -\left(\frac{qN_A}{\varepsilon_p \varepsilon_0} x + A\right), \qquad -x_p < x < 0 \tag{4.18}$$

To determine constant A we use boundary condition

$$E(-x_p) = 0 \tag{4.19}$$

One thus obtains

$$A = \frac{qN_A}{\varepsilon_p \varepsilon_0} x_p \tag{4.20}$$

From above, electric field in the p-region becomes

$$E(x) = -\frac{qN_A}{\varepsilon_p \varepsilon_0}(x + x_p), \qquad -x_p < x < 0 \tag{4.21}$$

Similarly, one finds for n-region

$$E(x) = -\frac{qN_D}{\varepsilon_n \varepsilon_0}(x - x_n), \qquad 0 < x < x_n \tag{4.22}$$

Plot of this situation, including charge distribution and electric field is shown in Fig. 4.4.

From above, one observes that electric field varies linearly in the depletion region. It reaches a minimum value at $x = 0$. The minimum value is

$$E_{min} = E(0) = -\frac{qN_A}{\varepsilon_p\varepsilon_0}x_p = -\frac{qN_D}{\varepsilon_n\varepsilon_0}x_n \qquad (4.23)$$

From above, one finds

$$\frac{N_A x_p}{\varepsilon_p} = \frac{N_D x_n}{\varepsilon_n} \qquad (4.24)$$

The above relation represents balance of charges in the depletion region, i.e. total positive charge equals to the total negative charge.

4.3.1.2 Solution for electrostatic potential

Electrostatic potential is defined as

$$E = -\frac{d\psi}{dx} \qquad (4.25)$$

Integrating the above equation gives the electrostatic potential. The total potential across the junction is equal to the difference between build-in potential and the applied voltage, i.e.

$$V_{bi} - V_{app} = \psi_n + \psi_p \qquad (4.26)$$

where ψ_n is the potential across the n-type region and ψ_p is the potential across the p-type region.

In p-region it obeys the Poisson equation

$$\frac{d\psi}{dx} = \frac{qN_A}{\varepsilon_p\varepsilon_0}(x + x_p), \qquad -x_p < x < 0 \qquad (4.27)$$

Integration of the above equation gives

$$\psi(x) = \frac{qN_A}{\varepsilon_p\varepsilon_0}x^2\frac{qN_A}{\varepsilon_p\varepsilon_0}x_p x + C$$

Constant C is determined by assuming that

$$\psi(x_p) = 0 \qquad (4.28)$$

and finally

$$\phi(x) = \frac{qN_A}{\varepsilon_p\varepsilon_0}(x_p - x)^2, \qquad -x_p < x < 0 \qquad (4.29)$$

Similar steps can be performed for a n-region. Different boundary condition is needed, namely

$$\phi(x_n) = V_0 \qquad (4.30)$$

For n-region one obtains

$$\phi(x) = V_0 - \frac{qN_D}{\varepsilon_n\varepsilon_0}(x_n - x)^2, \qquad 0 < x < x_n \qquad (4.31)$$

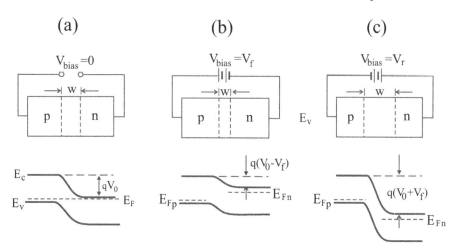

FIGURE 4.5
The effect of bias potential on a p-n junction. (a) equilibrium, (b) forward
bias, (c) reverse bias.

4.4   p-n Homojunction under Forward and Reverse Bias

Now, we will briefly summarize the behavior of p-n junction under external
voltage. We assume that an applied voltage V_{bias} appears entirely across the
transition region; see Fig. 4.5, from [3]. We shall take V_{bias} to be positive
when the external bias voltage is positive on the p side relative to the n side.
With the applied forward bias potential V_f, the electrostatic potential barrier
is lowered from the build-in value, or equilibrium contact potential value V_{bi}
to the smaller value $V_0 - V_f$. This lowering of the potential barrier occurs
because a forward bias (p positive with respect to n) raises the electrostatis
potential on the p side relative to the n side, as shown in Fig. 4.5. For a reverse
bias, the opposite occurs. The electrostatic potential of the p side is depressed
relative to the n side and the potential barrier at the junction becomes larger,
equal to $V_0 + V_r$. In the Fig. 4.5 we show change in the transition region width
w and also creation of a new Fermi levels E_{Fn} and E_{Fp}. Those are known
as quasi-Fermi levels. The difference in quasi-Fermi levels is related to the
applied potential bias V_{bias}.

4.5   Model of p-n Junction with Ohmic Contacts

As an illustration (an an example) of how ohmic contacts work we conduct here
a preliminary discussion of p-n junction assuming ohmic contacts. Created

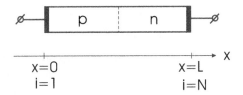

FIGURE 4.6
Schematic of a generic p-n diode.

model provides link between external bias voltage and carrier's concentrations at electrodes. Those results will find applications later-on.

The generic p-n diode which forms semiconductor laser is oriented as shown in Fig. 4.6 [4].

Distribution of all potentials ψ, ϕ_p, ϕ_n is shown in Fig. 4.7 [4]. Essentially, it may be assumed that ϕ_p and ϕ_n are constant outside depletion layer. We further choose $\phi_p = \phi_n = 0$ for $x = L$ (n-region). By assumption, n-region is sufficiently away from p-n junction.

External bias voltage V_{bias} is introduced by noting that for the thermal-equilibrium conditions one has

$$\phi_p = \phi_n = V_{bias} \qquad (4.32)$$

which holds for $x = 0$ (p-region) and is also sufficiently away from p-n junction.

Similarly, for $x = L$(n-region), one has

$$\phi_p = \phi_n = 0 \qquad (4.33)$$

Electrostatic potential (Boltzmann statistics)

$$p = n_{int} \exp \left[\frac{q}{kT} \left(\phi_p - \psi \right) \right] \qquad (4.34)$$

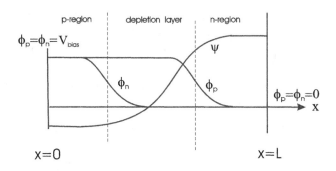

FIGURE 4.7
Distribution of potentials.

$$n = n_{int} \exp \left[\frac{q}{kT} \left(\psi - \phi_n \right) \right] \tag{4.35}$$

We assume that the device has ideal ohmic contacts at both electrodes where the excess carriers will be removed instantly, hence the charge neutrality will be maintained. In the thermal equilibrium, the boundary conditions for carrier concentrations are [4, 5] (no space charge)

$$C_{dop}(0) + p(0) - n(0) = 0 \tag{4.36}$$

$$C_{dop}(w) + p(w) - n(w) = 0 \tag{4.37}$$

and (assuming thermal equilibrium at both electrodes)

$$p(0) \cdot n(0) = n_i^2 \tag{4.38}$$

$$p(w) \cdot n(w) = n_{int}^2 \tag{4.39}$$

Concentrations are expressed as [4]

$$n = n_{int} e^{\Theta \psi} \tag{4.40}$$

$$p = n_{int} e^{\Theta (V - \psi)} \tag{4.41}$$

where V is an external voltage and $\Theta = 1/kT$. Solving the above equations gives

$$p(0) = -\frac{C_0(0)}{2} \left\{ 1 + \sqrt{1 + \left(\frac{2n_{int}}{C_0(0)} \right)^2} \right\}, \qquad n(0) = \frac{n_{int}^2}{p(0)} \tag{4.42}$$

$$n(w) = \frac{C_0(w)}{2} \left\{ 1 + \sqrt{1 + \left(\frac{2n_{int}}{C_{dop}(w)} \right)^2} \right\}, \qquad p(w) = \frac{n_{int}^2}{n(w)} \tag{4.43}$$

For $\frac{n_{int}}{C_{dop}(0)} \ll 1$ and $\frac{n_i}{C_{dop}(w)} \ll 1$ (which is always true except at very high temperatures), one has

$$p(0) \approx -C_{dop}(0), \qquad n(w) \approx C_{dop}(w) \tag{4.44}$$

In the above we selected plus sign as the carrier concentrations must be greater than or equal to zero.

The above formulas can be simplified by shifting potential ψ by ϕ_n. For that purpose introduce temporary potential ψ_t defined as $\psi_t = \psi - \phi_n$. Expressing the above carrier densities in terms of potential ψ_t gives

$$n = n_{int} \exp\left(\frac{q}{kT}\psi_t\right)$$

$$p = n_{int} \exp\left[\frac{q}{kT}(\phi_p - \psi_t - \phi_n)\right] = n_{int} \exp\left[\frac{q}{kT}(V_{bias} - \psi_t)\right]$$

where $V_{bias} = \phi_p - \phi_n$ is a bias voltage. Finally, replacing ψ_t by ψ gives

$$n = n_{int} \exp\left(\frac{q}{kT}\psi\right) \tag{4.45}$$

$$p = n_{int} \exp\left[\frac{q}{kT}(V_{bias} - \psi)\right] \tag{4.46}$$

which are the expressions used by Kurata [4]. Finally, substitute (6.15) and (6.15) into (6.13). Assume also homogeneous semiconductor with relative constant ε_r. Poisson equation becomes

$$\frac{d^2\psi}{dx^2} = -\frac{q n_{int}}{\varepsilon_0 \varepsilon_r}\left(e^{\frac{q}{kT}(V_{bias} - \psi)} - e^{\frac{q}{kT}\psi} + \frac{C_{dop}(x)}{n_{int}}\right) \tag{4.47}$$

Distribution of all potentials ψ, ϕ_p, ϕ_n is shown in Fig. 4.7. Both quasi-Fermi potentials ϕ_p (for holes) and ϕ_n (for electrons) may be regarded as constant outside the entire depletion region. This is an approximation, but can be considered as a definition of depletion layer.

4.6 p-i-n Diode

A p-i-n diode (or junction) is similar to an ordinary p-n junction. It contains undoped (intrinsic) region between p-type and n-type regions. Such structure is typical for a semiconductor laser.

To analyze this structure one must consider potential ψ_u in the intrinsic (undoped) region of thickness d. Therefore, Eq.(4.26) must be modified and it becomes [6]

$$\psi_n + \psi_p + \psi_u = V_{bi} - V_{app} \tag{4.48}$$

and

$$\psi_n = \frac{q N_D x_n^2}{2\varepsilon_s}, \quad \psi_p = \frac{q N_A x_p^2}{2\varepsilon_s}, \quad \psi_u = \frac{q N_A x_p d}{\varepsilon_s} \tag{4.49}$$

$$q N_A x_p = q N_D x_n \tag{4.50}$$

Here ψ_u is the potential across the middle undoped region of thickness d. From above equations one determines x_n [6]

$$x_n = \frac{\sqrt{d^2 + \frac{2\varepsilon_s}{q} \frac{N_D + N_A}{N_D N_A} (\psi_i - V_{app})} - d}{1 + \frac{N_D}{N_A}} \tag{4.51}$$

Using the above results one can determine potential across the structure

$$\psi(x) = \frac{qN_D}{2\varepsilon_s} (x + x_n)^2, \qquad -x_n < x < 0 \tag{4.52}$$

$$\psi(x) = -\psi_n - \frac{qN_D x_n}{\varepsilon_s}, \qquad 0 < x < d \tag{4.53}$$

$$\psi(x) = -(\phi_i - V_{app}) - \frac{qN_A}{2\varepsilon_s} (x - d - x_p)^2, \qquad d < x < d + x_p \tag{4.54}$$

We assumed that potential at $x = -x_n$ is zero.

4.7   Hetero p-n Junction

The hetero p-n junction is formed by two different semiconductors with opposite doping. It is in principle similar to a homojunction; now there exist discontinuities of parameters.

4.7.1   Formation of heterojunctions

Heterojunctions, typically p-n, are formed from two different semiconductor materials (e.g. *Si* and *SiGe*, or *GaAs* and *AlGaAs*) of different bandgap energies. Here we will only summarize fundamental properties of the basic types. More information can be found in [7–10].

Rules for forming equilibrium band diagram of the junction are

1. Fermi level is flat everywhere
2. far from junction the bulk like properties of the materials are recovered.

Now, we will discuss details of formation of $p - N$ and $n - P$ heterojunctions. We also analyze their behavior under forward and reverse bias voltage. Here, small letter refers to a semiconductor with smaller energy bandgap and capital letter refers to semiconductor with larger bandgap. We start with $p - N$ heterojunction.

4.7.1.1   Semiconductor $p - N$ heterojunction

Energy-band diagram of $p - N$ heterojunction are shown in Fig. 4.8, after [7]. Energy levels are shown before and after formation of an abrupt junction, as

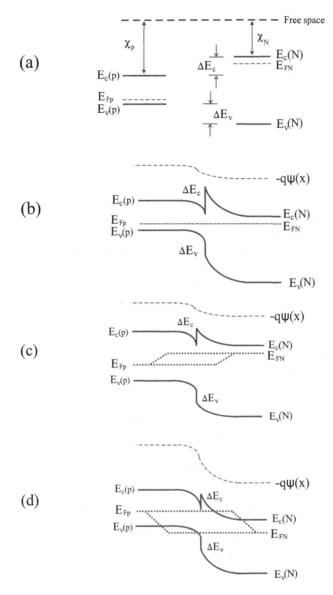

FIGURE 4.8

Energy-band diagram of an unbiased $p-N$ heterojunction before (a) and after (b) forming an abrupt junction. Energies under bias conditions are shown next: (c) corresponds to forward bias and (d) to reverse bias.

well as the situation under forward and reverse bias. Typical system might correspond to $p-GaAs$ and $N-Al_{0.3}Ga_{0.7}As$ materials. It is usually assumed that quasi-Fermi levels stay constant across the depletion region.

Fermi levels in each region are determined by the charge neutrality condition

$$n + N_A = p + N_D \tag{4.55}$$

Assuming Boltzmann distribution, charge densities on p side are [7]

$$
\begin{aligned}
n_p(x) &= N_{cp} \exp \left\{ [E_{FN}(x) - E_c(x)] / k_B T \right\} \\
p_p(x) &= N_{vp} \exp \left\{ [E_v(x) - E_{Fp}(x)] / k_B T \right\}
\end{aligned}
$$

On the N side

$$
\begin{aligned}
N_N(x) &= N_{CN} \exp \left\{ [E_{FN}(x) - E_c(x)] / k_B T \right\} \\
P_N(x) &= N_{VN} \exp \left\{ [E_v(x) - E_{Fp}(x)] / k_B T \right\}
\end{aligned}
$$

Combining the above equations one determines position-dependent quasi-Fermi levels. For details, and the final formulas consult [7].

With two regions in contact, the Fermi level will be constant across the junction under thermal equilibrium conditions and without bias voltage. As a result of redistribution of electrons and holes the build-in electric field will be created which in-turn prevents current flow in the crystal.

4.7.1.2   Semiconductor $n - P$ heterojunction

The formation of $n - P$ heterojunction is shown in Fig. 4.9, after [7]. Band diagram of anisotopic $p - N$ heterojuction before separation (a) and after contact (b) are shown as well as under external forward bias voltage (c) and reverse bias voltage (d).

As before, such junction is analyzed under the depletion approximation. One typically finds band diagrams as shown in Fig. 4.9.

4.7.1.3   Band diagram of a heterojunction p-n diode under flatband conditions

The flatband energy band diagram of a heterojunction p-n diode is shown in the Figure 4.10.

We assume that ΔE_c is positive if $E_{c,n} > E_{c,p}$ and ΔE_v is positive if $E_{v,n} < E_{v,p}$. The built-in potential ϕ_{bi} is defined as the difference between the Fermi levels in both the n-type and the p-type semiconductor. From the energy diagram we have

$$q\phi_{bi} = E_{F,n} - E_{F,p} = E_{F,n} - E_{c,n} + E_{c,n} + E_{c,p} - E_{F,p} \tag{4.56}$$

The built-in potential can be expressed as [6]

$$q\phi_{bi} = \Delta E_c + k_B T \ln \frac{n_{n0} N_{c,p}}{n_{p0} N_{c,n}} \tag{4.57}$$

One also has

$$q\phi_{bi} = -\Delta E_v + k_B T \ln \frac{p_{p0} N_{v,n}}{p_{n0} N_{v,p}} \tag{4.58}$$

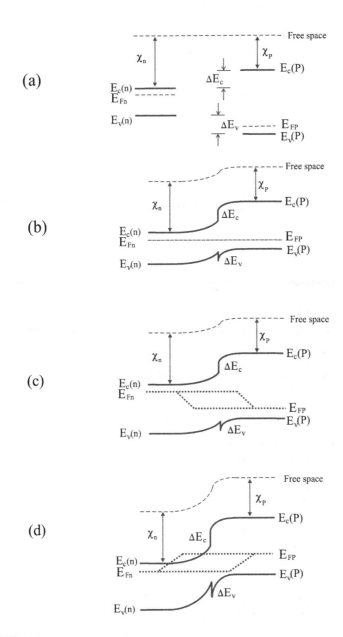

FIGURE 4.9
Energy band diagram of an unbiased n-P heterojunction at equilibrium: before
(a) and after (b) forming an abrupt junction. Energy band diagrams of a n-P
heterojunction under (c) forward bias and (d) reverse bias.

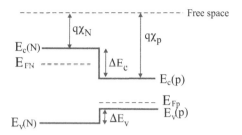

FIGURE 4.10
Flat-band energy band diagram of a p-n heterojunction

In tha above n_{in} and $n_{in,p}$ are the intrinsic carrier concentrations of the n and p-type region, respectively.

Bibliography

[1] R.A. Pierret. *Semiconductor Device Fundamentals*. Addison-Wesley, Reading, Massachusetts, 1996.

[2] S.M. Sze. *Physics of Semiconductor Devices*. Wiley, New York, 1969.

[3] B.G. Streetman and S. Banerjee. *Solid State Electronic Devices*. Prentice-Hall, Upper Saddle River, New Jersey, 2000.

[4] M. Kurata. *Numerical Analysis for Semiconductor Devices*. Lexington-Books, New York, 1982.

[5] M. Shur. *Physics of Semiconductor Devices*. Prentice Hall, Englewood Cliffs, NJ, 1990.

[6] Bart Jozef Van Zeghbroeck. *Principles of Semiconductor Devices*. Bart Van Zeghbroeck, 2007.

[7] S.-L. Chuang. *Physics of Photonics Devices*. Wiley, New York, 2009.

[8] B. Mroziewicz, M. Bugajski, and W. Nakwaski. *Physics of Semiconductor Lasers*. Polish Scientific Publishers, Warszawa, 1991.

[9] J.-M. Liu. *Photonic Devices*. Cambridge University Press, Cambridge, 2005.

[10] K.J. Ebeling. *Integrated Optoelectronics. Waveguide Optics, Photonics, Semiconductors*. Springer-Verlag, Berlin, 1993.

5

Electrical Processes

Here we summarize fundamental electrical processes taking place in p-n junction. They are parameterized by many material parameters. Their values, for important semiconductors are summarized in an Appendix. Optical processes are discussed in a separate chapter.

Material parameters are the most critical elements in determining quality of the simulations. They directly affect simulation results. For every meaningful simulation, proper choice of parameters is essential. In what follows we summarize all parameters, their values and/or compositional and temperature dependencies. The parameters are divided into several groups according to their common functionalities.

We start our discussion by describing general parameters and then summarizing their numerical values. We base our description on a phenomenological approach. When needed, microscopic approach will also be described. We combined parameters into the following groups:

1. general parameters (basic physical constants)

2. band structure parameters

3. doping parameters

4. carrier mobilities

5. recombination parameters

We start with basic physical constants.

5.1 Basic Physical Constants

The constants used for the simulations described here are summarized in Table 5.1. We provide typical symbol used in this book, its value and name.

DOI: 10.1201/9781003265849-5

TABLE 5.1

Basic physical constants.

Symbol	Value	Description
q	1.602×10^{-19} C	*electron's charge*
m_0	9.109×10^{31} kg	*electron's mass*
k_B	1.38×10^{-23} J/K	*Boltzmann constant*
ε_0	8.854×10^{-14} F/m	*dielectric constant*
h	6.626×10^{-34} J s	*Planck constant*
$\hbar$	1.054×10^{-34} J s	*Dirac constant*
T	300 K	*temperature*

5.2   Band Structure Parameters

Band structure parameters needed in the simulator developed here are summarized in Table 5.2. Their specific values will be discussed in later sections where we review practical materials. The band discontinuities ΔE_c and ΔE_v at an abrupt heterojunction are modelled using Anderson electron affinity rule [2]. Effective masses are determined by calculating a curvature at $k = 0$. Band gap energy is the difference between minimum of conduction and maximum of valence bands at Γ point (assuming direct transitions). Electron affinity is the difference between the bottom of the conduction band E_c and the vacuum energy level.

TABLE 5.2

Band parameters needed in simulations.

Symbol	Description
E_g	*band gap energy*
ΔE_c	*conduction band discontinuity*
ΔE_v	*valence band discontinuity*
χ_e	*electron affinity*
N_v	*effective densities of states for holes*
N_c	*effective densities of states for electrons*
m_c	*effective masses of electrons*
m_{hh}	*effective masses of heavy holes*
m_{lh}	*effective masses of light holes*
n_{int}	*intrinsic carrier density*

5.2.1 The effective densities of states

In the parabolic band model, the effective densities of states in 3D are given by formulas (2.10) and (2.18) which are summarized here again for completeness [3]

$$N_c = 2 \left(\frac{k_B T}{2\pi\hbar^2} \right)^{3/2} (m_e^*)^{3/2} = 2.51 \times 10^{19} \left(\frac{m_c^*}{m_0} \frac{T}{300} \right)^{3/2} cm^{-3} \qquad (5.1)$$

$$N_v = 2 \left(\frac{k_B T}{2\pi^2\hbar^2} \right)^{3/2} (m_v^*)^{3/2} = 2.51 \times 10^{19} \left(\frac{m_v^*}{m_0} \frac{T}{300} \right)^{3/2} cm^{-3} \qquad (5.2)$$

Here, N_c is the effective density of states in the conduction band and N_v is the effective density of states in the valence band.

5.3 Doping

Distribution of doping density which must be supplied by the user, enters the right side of Poisson's equation. It is used to determine the distribution of electrostatic potential. In the presence of dopants, which create impurity energy levels, the semiconductor becomes extrinsic. Two kinds of impurity atoms exist: those, which supply additional electrons (donors), and those, which accept electrons (acceptors). For bulk semiconductors in thermal equilibrium, the location of the Fermi level E_F is determined by the charge neutrality condition The procedure had been briefly discussed in Chapter 2.

In Fig. 5.1 we schematically show density of states and concentration of carriers for n-type, intrinsic and p-type semiconductors in thermal equilibrium (after Grundmann [4]). The locations of band edges for conduction and valence bands are shown as well as the positions of dopands (E_D) and acceptors (E_A).

The populations of acceptor and donor levels are given by the expressions [4]

$$N_A^- = \frac{N_A}{1 + g_A \exp\left(-\frac{E_F - E_A}{k_B T} \right)} \qquad (5.3)$$

$$N_D^+ = \frac{N_D}{1 + g_D \exp\left(\frac{E_F - E_D}{k_B T} \right)} \qquad (5.4)$$

Here g_A, g_D are the degeneracy factors. The acceptor concentration is denoted by N_A, the concentration of charged acceptors is N_A^-, E_A is the absolute acceptor energy. Similar terminology is introduced for donors.

Semiconductor lasers normally work at high density level of dopants. In this case, the quasi-Fermi levels are close to the impurity energy levels near the band edges (shallow impurity). Detailed doping models for specific materials

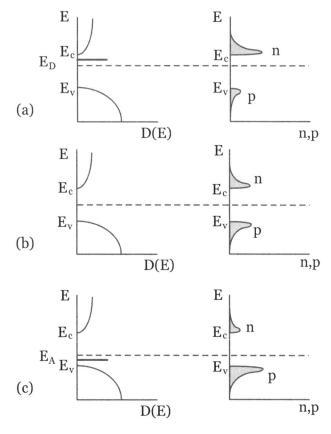

FIGURE 5.1
Density of states (left column) and carrier concentration (right column for (a)
n-type, (b) intrinsic and (c) p-type semiconductors in thermal equilibrium.

will be discussed in the later sections. List of typical doping parameters which
are needed is provided in Table 5.3.

TABLE 5.3
Doping parameters.

Symbol	Description
N_D	*donor impurity concentration*
N_A	*acceptor impurity concentration*
N_{res}	*residual doping*

TABLE 5.4

Mobility parameters.

Symbol	Description
μ_n	*mobility of electrons*
μ_p	*mobility of holes*

5.4 Carrier Mobilities

Mobility plays important role in device's simulation. Carrier mobilities μ_n and μ_p account for scattering mechanisms in electrical transport. Mobilities of carriers in a device directly determine the operation speed of the device. They also affect the leakage current due to diffusion over the barrier. List of mobility parameters is provided in Table 5.4. Some of the typical values of mobilities for various semiconductors are given in Table 5.5 (from [4]).

Carrier mobilities of semiconductor crystal are described using models ranging from the simplest Drude model to those including various kinds of scattering mechanisms. Various mobility models can be found in literature. In general, the mobility is a function of the dopant densities and the electrical field. Various mobility models can be combined. The total mobility μ then results from the mobilities μ_i of the individual models according to Mathiessen's rule [5]

$$\frac{1}{\mu} = \sum_i \frac{1}{\mu_i} \qquad (5.5)$$

Popular models of mobility are [6]: constant mobility, Croslight [7, 8], Minimos [9], Arora [10], Masetti [11]. We will discuss them in detail now.

TABLE 5.5

Mobilities (in cm^2/Vs) for electrons and holes at room temperature for various semiconductors.

	μ_n	μ_p
Si	1300	500
Ge	4500	3500
GaAs	8800	400
GaN	300	180
InSb	77000	750
InAs	33000	460
InP	4600	150
ZnO	230	8

- Constant mobility is given by the following formula [10]

$$\mu_{const} = \mu_L \left(\frac{T}{300}\right)^{-\varsigma} \tag{5.6}$$

The constant mobility is independent of the impurity concentration $N_i = N_A + N_D$. It depends only on the two material dependent parameters μ_L and ς. For example, for lightly doped n-type silicon within the $200 - 500K$ range of temperatures the fit to experimental data gives [10]

$$\mu = 8.56 \times 10^8 T^{-2.33} \tag{5.7}$$

where T is the temperature in Kelvins.

- Minimos [9]. The model of mobility implemented in Minimos is

$$\mu(x) = \mu_{min} + \frac{\mu_L - \mu_{min}}{1 + \left(\frac{N_i}{C_{ref}}\right)^\alpha} \tag{5.8}$$

where

$$\mu_{min} = \begin{cases} \mu_{min,300} \left(\frac{T}{300}\right)^{\gamma_1} & \text{for } T \geq T_{switch} \\ \mu_{min,300} \left(\frac{2}{3}\right)^{\gamma_1} \left(\frac{T}{300}\right)^{\gamma_2} & \text{for } T < T_{switch} \end{cases} \tag{5.9}$$

$$C_{ref} = C_{ref,300} \left(\frac{T}{300}\right)^{\gamma_3} \tag{5.10}$$

$$\alpha = \alpha_{300} \left(\frac{T}{300}\right)^{\gamma_4} \tag{5.11}$$

The Minimos mobility model depends on the following nine material dependent parameters: $\mu_L, \mu_{min,300}, \gamma_1, \gamma_2, \gamma_3, \gamma_4, T_{switch}, C_{ref,300}, \alpha_{300}$.

- Arora [10]. The mobility model proposed by Arora et al is

$$\mu(x) = \mu_{min} + \frac{\mu_d}{1 + \left(\frac{N_i}{N_0}\right)^{A^*}} \tag{5.12}$$

where

$$\mu_{min} = A_{min} \left(\frac{T}{300}\right)^{\alpha_m}, \quad \mu_d = A_d \left(\frac{T}{300}\right)^{\alpha_d} \tag{5.13}$$

$$N_0 = A_N \left(\frac{T}{300}\right)^{\alpha_N}, \quad A^* = A_\alpha \left(\frac{T}{300}\right)^{\alpha_\alpha} \tag{5.14}$$

The Arora mobility model depends on the following eigth material dependent parameters: $A_{min}, \alpha_m, A_d, \alpha_d, A_N, \alpha_N, A_\alpha, \alpha_\alpha$.

- Masetti [11]. The mobility model proposed by Masetti et al is

$$\mu = \mu_{\min 1} \exp\left(-\frac{P_C}{N_i}\right) + \frac{\mu_{const} - \mu_{\min 2}}{1 + \left(\frac{N_i}{C_r}\right)^\alpha} - \frac{\mu_1}{1 + \left(\frac{C_s}{N_i}\right)^\alpha}$$

The Masetti mobility model depends on the following eight material dependent parameters: $\mu_{\min 1}, P_C, \mu_{\min 2}, C_r, \alpha, \mu_1, C_s, \beta$. In addition, the mobility μ_{const} from the constant mobility model enters the Masetti mobility model.

- Croslight. In the LASTIP manual [12] there is an extensive summary of the dependence of mobility on the electric field:

  (a) In the simplest model of mobilities one uses constant mobility μ_{0n} and μ_{0p} for electrons and holes, respectively, throughout each material region in the device.

  (b) More complicated model contains two-piece dependence

$$\begin{aligned}
\mu_n &= \mu_{0n}, \quad \text{for } E < E_{0n} \\
\mu_n &= v_{sn}/E, \quad \text{for } E \geq E_{0n} \qquad (5.15) \\
v_{sn} &= \mu_{0n} E_{0n}
\end{aligned}$$

  for the electron mobility. E_{0n} is a threshold field beyond which the electron velocity saturates to a constant. Similar expression can be defined for holes

$$\begin{aligned}
\mu_p &= \mu_{0p}, \quad \text{for } E < E_{0p} \\
\mu_p &= v_{sp}/E, \quad \text{for } E \geq E_{0p} \qquad (5.16) \\
v_{sp} &= \mu_{0p} E_{0p}
\end{aligned}$$

  (c) The electric field dependence of mobilities can be described by the following equations

$$\mu_n = \frac{\mu_{n0}}{\left[1 + \left(\frac{\mu_{n0} E}{v_{sn}}\right)^{\beta_n}\right]^{1/\beta_n}} \qquad (5.17)$$

$$\mu_p = \frac{\mu_{p0}}{\left[1 + \left(\frac{\mu_{p0} E}{v_{sp}}\right)^{\beta_p}\right]^{1/\beta_p}} \qquad (5.18)$$

  Here E, v_{sn}, v_{sp} denote electric field and saturation electron and hole velocities. Also μ_{n0} and μ_{p0} are low-field mobilities. In this model velocities increase linearly with electric field and saturate at the high field limit. Field dependence is controlled by the parameters β_n and β_p, for electrons and holes, respectively.

(d)   The effect of impurity dependence of mobilities at low electric fields is described by the following empirical expression

$$\mu_{n0} = \mu_{1n} + \frac{\mu_{2n} - \mu_{1n}}{\left(\frac{N_D + N_A + \sum_j N_{tj}}{N_{rn}}\right)^{\alpha_n}} \qquad (5.19)$$

$$\mu_{n0} = \mu_{1n} + \frac{\mu_{2n} - \mu_{1n}}{\left(\frac{N_D + N_A + \sum_j N_{tj}}{N_{rn}}\right)^{\alpha_n}} \qquad (5.20)$$

where parameters are obtained by fitting to experimental data. Here N_{tj} is the concentration of deep level trap (of the j-th level) [7].

- Silvaco. ATLAS II User's Manual (1993) [13]. They provided several models of mobility which quarantee realistic results. The considered effects on mobility were: doping dependence, temperature dependence and electric field dependence.

5.5   Recombination

Main recombination processes considered in this book involves: spontaneous recombination, stimulated recombination, Shockley-Read-Hall (SRH) generation-recombination and Auger recombination. Parameters describing those processes are summarized in Table 5.6. In Chapter 2, we briefly discussed basic transition processes. In the following we will discuss both radiative and non-radiative processes in more detail, provide practical working expressions describing those processes as well as the values of the coefficients appearing in those expressions.

Four types of recombination processes are included in our model: spontaneous recombination, stimulated recombination, Auger non-radiative

TABLE 5.6

Recombination parameters.

Symbol	Description
B_0	*spontaneous recombination coefficient*
B_1	*spontaneous recombination coefficient*
C_n	*Auger recombination coefficient*
C_p	*Auger recombination coefficient*
τ_{n0}	*SRH carrier life-time (electrons)*
τ_{p0}	*SRH carrier life-time (holes)*

recombination, and Shockley-Read-Hall (SRH) non-radiative recombination. The total recombination rate is thus

$$R = R_{spon} + R_{stim} + R_{Auger} + R_{SRH} \tag{5.21}$$

R_{spon} describes spontaneous emission, R_{stim} is responsible for stimulated emission, R_{Auger} describes Auger processes, and finally R_{SRH} describes Shockley-Read-Hall recombination. Those processes will now be described in detail.

5.5.1 Spontaneous recombination

The spontaneous recombination rate is given by [14]

$$R_{spon} = B(np - n_{int}p_{int}) \tag{5.22}$$

where B is the spontaneous recombination coefficient and n_{int} and p_{int} are the intrinsic densities. Coefficient B has the following form

$$B = B_0 - B_1 \min(n, p) \tag{5.23}$$

For *AlGaAs* we use simpler model with $B = const.$

5.5.2 Stimulated recombination

General expression for stimulated recombination for arbitrary number of modes was provided by Ohtoshi et al [14]. In the case of a single mode, the stimulated recombination rate is [3]

$$R_{stim} = g(n, p)v_g S|E(x)|^2 \tag{5.24}$$

where $|E(x)|^2$ is the electric field mode profile, S the photon density, $v_g = c/n$ the group velocity and $g(n, p)$ the local optical gain. The optical gain considered here is introduced in a phenomenological way.

5.5.3 Shockley-Read-Hall (SRH) generation-recombination

The Shockley-Read-Hall (SRH) generation-recombination (G-R) involves four processes:

a) electron capture

b) hole capture

c) electron emission

d) hole emission.

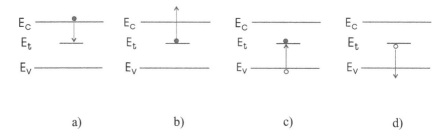

a)   b)   c)   d)

FIGURE 5.2
The energy band diagrams for the SRH G-R processes. The processes are: a) electron capture, b) electron emission, c) hole capture, d) hole emission.

Those processes involve special sites within the semiconductors known as G-R centers or traps. Those centers have energies between the conduction and valence bands. The transitions are primarily two-step processes. The situation is shown schematically in Fig. 5.2. We introduce G-R centers which are characterized by E_t the energy level of a trap and N_t the concentrations of traps. We also introduce capture coefficients for electrons and holes. We now discuss the four processes separately:

Electron capture

It occurs when a conduction band electron is trapped by a vacant G-R center. Consequently, the G-R center becomes occupied. The captured electron can either fall to the valence band and annihilate a hole (hole capture) or move back to conduction band (electron emission). In both cases the G-R center is vacanting. Mathematically, the process of electron capture is described by the equation

$$R_n = c_n n N_t \left(1 - f_t\right) \tag{5.25}$$

where c_n the proportionality factor is called capture coefficient. It has the dimension $cm^3 s^{-1}$ and describes the capture capability of an unoccupied deep center. Density of electrons is represented by n and f_t is the occupation probability of the trap. Consequently, $(1 - f_t)$ describes the probability that the trap is empty.

Hole capture

This process is similar to the above electron's capture. Here holes are captured by occupied traps. The recombination rate of hole's capture is

$$R_p = c_p p N_t f_t \tag{5.26}$$

Here c_p is the capture coefficient for holes.

Electron emission

As stated above, electron's emission occurs when electron vacate the G-R center. The generation rate for electrons is

$$G_n = e_n N_t f_t \tag{5.27}$$

where e_n is the emission coefficient of electrons and $N_t f_t \equiv n_t$ is the density of the traps which are occupied by the electrons.

Hole emission

Hole emission occurs when a valence band electron is trapped by a vacant G-R center. Consequently, the G-R center becomes occupied and a hole is produced in the valence band. This electron can either fall back to the valence band and annihilate a hole (hole capture) or move to the conduction band (electron emission). Again, the occupied G-R center becomes vacant.

The generation rate of holes is therefore

$$G_p = e_p N_t \left(1 - f_t\right) \tag{5.28}$$

where e_p is the emission coefficient for the holes.

If hole emission is followed by electron emission, then generation occurs and an electron-hole pair is created. If electron capture is followed by hole capture, then recombination occurs and an electron-hole pair is destroyed.

In order to obtain a mathematical expression for SRH net recombination rate, consider a system in thermal equilibrium. Based on the principle of detailed balance, the net G-R of electrons and holes should be zero. Therefore

$$R_n - G_n = c_n n_0 N_t \left(1 - f_{t0}\right) - e_n N_t f_{t0} = 0 \tag{5.29}$$

$$R_p - G_p = c_p p_0 N_t f_{t0} - e_p N_t \left(1 - f_{t0}\right) = 0 \tag{5.30}$$

where the subscript 0 signifies the thermal equilibrium values.

Solving the above equations, one finds

$$e_n = c_n n_0 \frac{1 - f_{t0}}{f_{t0}} \equiv c_n n_1 \tag{5.31}$$

$$e_p = c_p p_0 \frac{f_{t0}}{1 - f_{t0}} \equiv c_p p_1 \tag{5.32}$$

Combining Eqs.(5.25) and (5.27) and using Eq.(5.31) we have

$$\begin{aligned} R_n - G_n &= c_n n N_t \left(1 - f_t\right) - e_n N_t f_t \\ &= c_n n N_t \left(1 - f_t\right) - c_n n_1 N_t f_t \\ &= c_n N_t \left[n \left(1 - f_t\right) - n_1 f_t\right] \end{aligned} \tag{5.33}$$

Similarly

$$R_p - G_p = c_p N_t \left[p f_t - p_1 \left(1 - f_t\right)\right]$$

In equilibrium, when two pair of processes (creation and annihilation of electron-hole pair) are balanced, i.e. $R_n - G_n = R_p - G_p$, we have the relation

$$c_n N_t \left[n \left(1 - f_t\right) - n_1 f_t\right] = c_p N_t \left[p f_t - p_1 \left(1 - f_t\right)\right]$$

We can use the above relation to determine the fraction of occupied traps f_t as

$$f_t = \frac{c_n n + c_p p_1}{c_n \left(n + n_1\right) + c_p \left(p + p_1\right)}$$

Substitute the last result into (5.33) and after several algebraic steps one obtains

$$R_n - G_n = N_t \frac{c_n c_p np - e_n e_p}{c_n n + e_n + c_p p + e_p} \tag{5.34}$$

where $e_n = c_n n_1$ and $e_p = c_p p_1$. Introducing electron and hole lifetimes

$$\tau_n = \frac{1}{c_n N_t} \tag{5.35}$$

$$\tau_p = \frac{1}{c_p N_t} \tag{5.36}$$

Eq. (5.34) can be expressed in the final form as the SRH recombination rate

$$R_{SRH} = R_n - G_n = \frac{np - n_{int} p_{int}}{\tau_p(n + n_{int}) + \tau_n(p + p_{int})} \tag{5.37}$$

5.5.4 Auger recombination

At long wavelength (smaller bandgaps) non-radiative Auger recombination in which an electron and hole recombine and transfer their energies to another carriers is recognized as the main mechanism.

Those processes are illustrated in Fig. 5.3. For semiconductors, the key element is that at least three particles (two electrons and one hole, or two holes and one electron) and four energy states are involved. More generally, the transitions may be band-to-band, may involve photon assistance or may involve impurities.

Main Auger processes are [1, 3]: CHCC, CHLH, CHHS, CHHH. The notation used is explained as follows: C - refers to conduction band, H - to heavy-hole band, L - to light-hole band, and S - to spin-split-off band.

As an example, the CHCC process involves a transition from the conduction band to the heavy hole band at a different momentum and a subsequent transfer of the excess energy to a low lying conduction electron which is excited to a higher energy within the conduction band. The CHHS process involves one electron, two heavy holes and a split-off-band hole. CHHL is similar to CHHS except that it involves a light hole. Microscopic calculations of Auger processes were reviewed by Agrawal and Dutta [1]. We will now briefly describe CHCC process and refer to a book by Agrawal and Dutta [1] for description of the remaining processes.

Simplified illustration of these processes is shown in Fig. 5.4. We now discuss them in detail.

5.5.4.1 CHCC process

Electron in state a recombines with a heavy-hole at b, and both the energy and momentum conservation are taken up by an electron which moves from state c to state d. The above Auger band-to-band processes will now be represented

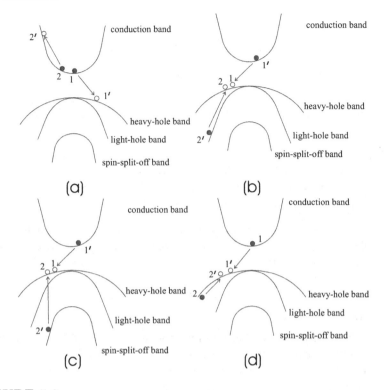

FIGURE 5.3
Band-to-band Auger recombination processes. (a) CHCC, (b) CHLH, (c) CHHS, (d) CHHH processes.

on the energy band diagrams as shown in Fig. 5.4. The four possible processes are [3]:

a) **Electron capture**

In this process an electron in the conduction band recombines with a hole in the valence band and releases its energy to a nearby electron. This process destroys an electron-hole pair. The recombination rate in this process is described by the phenomenological equation

$$R_n = c_n^{Au} n^2 p \qquad (5.38)$$

b) **Electron emission**

In this process an electron jumps from the valence band to the conduction band. This process creates an electron-hole pair and it is described by

$$G_n = e_n^{Au} n \qquad (5.39)$$

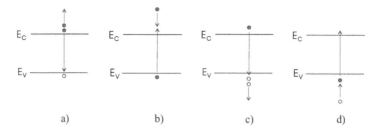

FIGURE 5.4
The energy band diagrams for the band-to-band Auger processes. a) electron capture, b) electron emission, c) hole capture, d) hole emission.

c) **Hole capture**

An electron from the conduction band recombines with a hole in the valence band. The released energy is taken by a nearby hole. This process destroys an electron-hole pair. Phenomenologically, it is described by

$$R_n = c_n^{Au} n^2 p \tag{5.40}$$

d) **Hole emission**

In this process an electron from the valence band jumps to the conduction band due to collision with a hole in the valence band. The generation rate for this process is

$$R_n = c_n^{Au} n^2 p \tag{5.41}$$

Total Auger recombination rate

In thermal equilibrium there is no net generation/recombination. Therefore processes a) and b) and also processes c) and d) balance each other. One therefore finds

$$R_n - G_n = c_n^{Au} n_0^2 p_0 - e_n^{Au} n_0 = 0 \tag{5.42}$$

and

$$R_p - G_p = c_p^{Au} n_0 p_0^2 - e_p^{Au} p_0 = 0 \tag{5.43}$$

From above one finds relations between phenomenological coefficients

$$e_n^{Au} = c_n^{Au} n_0 p_0 = c_n^{Au} n_{int}^2 \tag{5.44}$$

and

$$e_p^{Au} = c_p^{Au} n_0 p_0 = c_p^{Au} n_{int}^2 \tag{5.45}$$

Using the above results, the net total Auger recombination rate is

$$R_{Auger} = (C_n n + C_p p)(np - n_{int} p_{int}) \tag{5.46}$$

where C_n and C_p are the Auger recombination coefficients, respectively, associated with the CHCC and the CHSH Auger process. Auger coefficients C_n and C_p depends on the type of material simulated.

We assume that Eq.(13.17) is valid for both bulk and QW regions and consider coefficients C_n and C_p as empirical parameters.

Bibliography

[1] G.P. Agrawal and N.K. Dutta. Semiconductor Lasers. Second Edition. *Kluwer Academic Publishers*, Boston, 2000.

[2] R.L. Anderson. Experiments on Ge-GaAs heterojunctions. *Solid State Electron.*, 5:341–351, 1962.

[3] S.-L. Chuang. *Physics of Photonics Devices*. Wiley, New York, 2009.

[4] M. Grundmann. *The Physics of Semiconductors. An Introduction Including Devices and Nanophysics*. Springer, Berlin, 2006.

[5] N.W. Ashcroft and N.D. Mermin. *Solid State Physics*. Holt-Saunders, New York, 1976.

[6] Tobias Zibold. *Semiconductor based quantum information devices: Theory and simulations*. PhD thesis, Technische Universitat Munchen, 2007.

[7] Z.-M. Li. Two-dimensional numerical simulation of semiconductor lasers. In W.P. Huang, editor, *Electromagnetic Waves, Methods for Modeling and Simulation of Guided-Wave Optoelectronic Devices: Part II:Waves and Interactions*, volume PIER 11, pages 301–344. 1995.

[8] For actual update, visit company at www.crosslight.com.

[9] S. Selberherr, A. Schuetz, and H.W. Poetzl. Minimos-a two-dimensional mos transistor analyzer. *IEEE J. Solid-State Circuits*, 15:605–615, 1980.

[10] N.D. Arora, J.R. Hauser, and D.J. Roulston. Electron and hole mobilities in silicon as a function of concentration and temperature. *IEEE Trans. Electron Devices*, 29:292–295, 1982.

[11] G. Masetti, M. Severi, and S. Solmi. Modeling of carrier mobility against carrier concentration in arsenic-, phosphorus-, and boron-doped silicon. *IEEE Trans. Electron Devices*, 30:764–769, 1983.

[12] https://crosslight.com/products/lastip/.

[13] 1993. ATLAS II. User's Manual. Version 1.0.

[14] T. Ohtoshi, K. Yamaguchi, C. Nagaoka, T. Uda, Y. Murayama, and N. Chinone. A two-dimensional device simulator of semiconductor lasers. *Solid State Electron.*, 30:627–638, 1987.

6

Poisson Equation

Starting with this chapter we proceed with the code development using Matlab platform. Our final goal is to create a simple simulator of one-dimensional (1D) semiconductor laser and we reach it in several steps of increased complexity. As a laser operates as p-n junction we initialize our work with numerical modeling of a p-n junction.

The simplest approach to model p-n junction is based on using Poisson equation. Such approach describes equilibrium case. We use it to model homogeneous junction and also heterojunction. After those steps we generalize it by using drift-diffusion based approach.

The drift-diffusion method is based on three coupled differential equations: Poisson equation and current continuity equations for electrons and holes. It is a standard approach in computational electronics. Comprehensive summary of computational electronics, including the drift-diffusion model as well as its extensions is described in a book by Vasileska et al [1].The central role is played by Poisson equation. It links charge distribution within a sample with electric potential. The charge distribution is influenced by the non-equilibrium behavior within the device.

In some situations the nonlinear Poisson equation can be solved separately as a stand alone equation. The simplest application of such a result involves one-dimensional model of pn diode [2, 3]. More involved approaches of solving Poisson's equation for semiconductor heterostructures and also the generalization to two-dimensional cases were reported by Mayergoyz [4], Krowne [5] and Jozwikowska [6]. The obtained solutions applies to problem when one deals with devices in thermal equilibrium or are part of Gummel's algorithm or its modifications.

We adopt the convention that junction in equilibrium is described solely using Poisson equation. Doping is predefined and densities of electrons and holes are given in terms of electrostatic field (Boltzmann limit). In this chapter we use build-in Matlab functions for solving algebraic equations. We first discuss homostructures and later heterostructures.

Keep in mind that the code here was developed for didactic purposes and as such still can be improved and optimized.

DOI: 10.1201/9781003265849-6

6.1 Simple Poisson Equation

Typical form of Poisson equation used in semiconductor physics was provided earlier. Assume first that the right-hand side (RHS) does not depend on potential function $u(x)$. One therefore arrives at the following equation

$$\frac{d^2 u(x)}{dx^2} = g(x) \tag{6.1}$$

and $a < x < b$ and $g(x)$ is some known function. Dirichlet boundary conditions are: $u(a) = u_a$ and $u(b) = u_b$ where u_a and u_b are known values. To solve the problem, divide the computational domain $a < x < b$ into equal segments defined by grid points, as shown in Fig. 6.1

$$x_i = a + i \frac{b-a}{N+1}, \quad \text{for } i = 1, 2, ...N \tag{6.2}$$

The boundaries correspond to $i = 0$ and $i = N + 1$.

The discretization of the differential term $\frac{d^2 u(x)}{dx^2}$ is done using a second-order central difference scheme, as

$$\frac{d^2 u(x_i)}{dx^2} = \frac{u_{i-1} - 2u_i + u_{i+1}}{(\Delta x)^2} + O(\Delta x)^2 \tag{6.3}$$

Here $\Delta x = \frac{b-a}{N+1}$ and $u_i = u(x_i)$. The discretized version of Poisson equation is therefore

$$\frac{u_{i-1} - 2u_i + u_{i+1}}{(\Delta x)^2} = g_i \tag{6.4}$$

or

$$u_{i-1} - 2u_i + u_{i+1} = (\Delta x)^2 g_i \quad \text{for } i = 1, 2, ...N \tag{6.5}$$

where $g_i = g(x_i)$. From equation (6.5) one finds the following equations corresponding to a particular value of i

$$
\begin{aligned}
i = 1 \qquad & u_0 - 2u_1 + u_2 = (\Delta x)^2 \cdot g_1 \\
i = 2 \qquad & u_1 - 2u_2 + u_3 = (\Delta x)^2 \cdot g_2 \\
& \cdots \\
i = N-1 \qquad & u_{N-2} - 2u_{N-1} + u_N = (\Delta x)^2 \cdot g_{N-1} \\
i = N \qquad & u_{N-1} - 2u_N + u_{N+1} = (\Delta x)^2 \cdot g_N
\end{aligned}
\tag{6.6}
$$

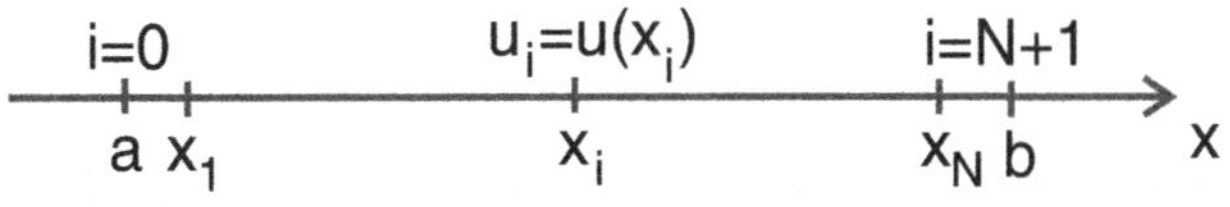

FIGURE 6.1
Notation for discretization of Poisson equation.

Observe that $u_0 = u_a$ and $u_{N+1} = u_b$ are known quantities. Therefore the above system takes the following form where the unknowns are on the left-hand side (LHS) and known values are on the RHS

$$
\begin{aligned}
-2u_1 \quad +u_2 & = (\Delta x)^2 \cdot g_1 - u_a \\
u_1 \quad -2u_2 \quad +u_3 & = (\Delta x)^2 \cdot g_2 \\
& \quad \cdots \\
u_{N-2} \quad -2u_{N-1} \quad +u_N & = (\Delta x)^2 \cdot g_{N-1} \\
u_{N-1} \quad -2u_N & = (\Delta x)^2 \cdot g_N - u_b
\end{aligned}
\tag{6.7}
$$

The above equations can be formulated in a matrix form. For this, define matrix $\overleftrightarrow{A}$

$$
\overleftrightarrow{A} =
\begin{bmatrix}
-2 & 1 & 0 & & & 0 & 0 \\
1 & -2 & 1 & & & & \\
0 & 1 & -2 & 1 & & & \\
& & & \cdots & & & \\
& & & & 1 & -2 & 1 \\
0 & & 0 & & & 0 & 1 & -2
\end{bmatrix}
\tag{6.8}
$$

and vectors $\vec{u}$ and $\vec{g}$ as

$$
\vec{u} =
\begin{bmatrix}
u_1 \\ u_2 \\ \cdots \\ u_N
\end{bmatrix}
\qquad
\vec{g} =
\begin{bmatrix}
(\Delta x)^2 \cdot g_1 - u_a \\
(\Delta x)^2 \cdot g_2 \\
\cdots \\
(\Delta x)^2 \cdot g_{N-1} \\
(\Delta x)^2 \cdot g_N - u_b
\end{bmatrix}
\tag{6.9}
$$

The discretized system of equations 6.7 can be written as

$$
\overleftrightarrow{A} \cdot \vec{u} = \vec{g}
\tag{6.10}
$$

After general discussion we introduce an example.

Example

Solve the following Poisson's equation along with the boundary conditions defined in the interval $0 \le x \le 1$

$$
\begin{aligned}
u'' & = -x^2 \\
u(0) & = 0 \\
u(1) & = 0
\end{aligned}
\tag{6.11}
$$

Exact solution is found as follows: first, write it as

$$
\frac{d}{dx}\left(\frac{du}{dx}\right) = -x^2
$$

Integrate

$$\frac{du}{dx} = -\frac{1}{3}x^3 + A$$

and

$$u(x) = -\frac{1}{3} \cdot \frac{1}{4}x^4 + Ax + B$$

The exact solution is therefore

$$u_{exact}(x) = -\frac{1}{12}x^4 + Ax + B$$

where A and B are constants. Applying boundary conditions we determine constants

$$u_{exact}(0) = 0 = B \quad \rightarrow \quad B = 0$$

$$u_{exact}(1) = 0 = -\frac{1}{12} + A \quad \rightarrow \quad A = \frac{1}{12}$$

Finally, the complete solution is

$$u_{exact}(x) = -\frac{1}{12}x^4 + \frac{1}{12}x \tag{6.12}$$

Exact solution (which is still obtained using numerical methods) is plotted in Fig. (6.2) and compared with our numerical solution obtained using Eq.(6.10). Observe how mesh points are generated. In our simulations boundaries are treated separately and mesh points are only generated inside the structure. a and b are coordinates of contacts.

Numerical code is provided in Appendix, Listing 6.A.1. In the developed code boundary conditions are naturally accounted for when we create mesh points.

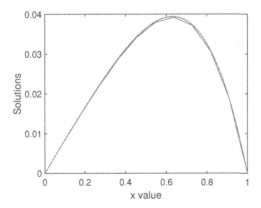

FIGURE 6.2
Comparison of the exact and numerical results. Number of mesh points N=10. Solid line – exact solution, dotted – numerical result.

After this introductory section we start more serious analysis and applications of Poisson equation using p-n junction as an example.

6.2   p-n Diode in Equilibrium

We start with a discussion of a p-n junction in equilibrium. Our approach here is based on numerical solution of Poisson equation for electrostatic potential assuming specific dependencies of electron and hole densities on the potential but without considering their dynamics.

The generic p-n diode has been introduced before. Here we show it again, as shown in Fig. 6.3. Observe that we have freedom of how we define mesh points at contacts. Here, first and last mesh points define contacts.

In one dimension, Poisson equation takes the form

$$\frac{d}{dx} \cdot \left(\varepsilon \frac{d\psi}{dx}\right) = -q(p - n + C_{dop}(x)) \tag{6.13}$$

where

$$C_{dop}(x) = N_D^+(x) - N_A^-(x) \tag{6.14}$$

and $\varepsilon = \varepsilon_0 \varepsilon_r(x)$, where $\varepsilon_r = \overline{n}^2(x)$ is the relative dielectric constant at local point and $\overline{n}(x)$ is a local refractive index at position x. In the Boltzmann limit, carrier densities are expressed as (relations established previously)

$$n = n_{int} \exp\left(\frac{q}{kT}\psi\right) \tag{6.15}$$

and

$$p = n_{int} \exp\left[\frac{q}{kT}\left(V_{bias} - \psi\right)\right] \tag{6.16}$$

or, in terms of quasi-Fermi potentials ϕ_p and ϕ_n as

$$n = n_{int} \exp\left(\frac{\psi - \phi_n}{V_T}\right) \tag{6.17}$$

FIGURE 6.3
Schematic of a generic p-n diode.

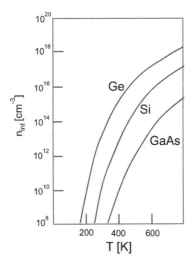

FIGURE 6.4
Temperature dependence of intrinsic carrier concentration for several semi-conductors.

and

$$p = n_{int} \exp \left(\frac{\phi_p - \psi}{V_T} \right) \tag{6.18}$$

where V_{bias} is an external bias voltage, ψ is electrostatic potential, q is electron's charge, k_B is Boltzmann constant and T is the absolute temperature.

The intrinsic concentration n_{int} corresponds to the concentration of electrons or holes in an undoped (pure crystal), i.e. intrinsic semiconductor. In such a semiconductor $n = p = n_{int}$ which is therefore called the intrinsic concentration. We assume here

$$n_{int} = 1.4 \times 10^{16} m^{-3} \tag{6.19}$$

Intrinsic carrier concentrations of Ge, Si and GaAs as a function of temperature are shown in Fig. 6.4. The values of fundamental constants are summarized in Table 6.1.

TABLE 6.1
The values of fundamental constants.

Description	Symbol	Value
Permittivity constant	ε_0	$8.85 \times 10^{-12} \frac{C^2}{N \cdot m^2}$
Elementary charge	q	$1.6 \times 10^{-19} C$
Boltzmann constant	k_B	$1.38 \times 10^{-23} \frac{J}{K}$

TABLE 6.2

Definitions of scaling variables.

Name	Quantity	Definition of scaling
distance	x	$x' = \frac{x}{x_0}$
potential	ψ	$\psi' = \frac{\psi}{V_T}$
concentrations	n, p, C_{dop}	$n' = \frac{n}{C_0}, p' = \frac{p}{C_0}, C'_{dop} = \frac{C_{dop}}{C_0},$
parameter (linear junction)	m	$m' = m\frac{x_0}{C_0}$

In what follows we will concentrate on numerical solution of 1D Poisson equation. For that purpose we apply scaling for all variables. Full appreciation of scaling will be evident in a later chapter where we analyze electric currents. The problem is that the quantities analyzed change by many orders of magnitude over short distances. Numerical implementation will be discussed only after we formulate boundary conditions and establish the choice of trial values. Before that, we will review scaling of Poisson equation.

6.3   Scaling of Poisson Equation

For the scaling of the above equations we employ definitions as specified in Table 6.2 and simple scaling factors are summarized in Table 6.3. After scaling Poisson equation takes the form

$$\frac{d}{dx'} \cdot (\varepsilon_r \frac{d\psi'}{dx'}) = -(p' - n' + C'_{dop}(x')) \tag{6.20}$$

The densities are

$$n'(x) = e^{\psi'(x)} \tag{6.21}$$

$$p'(x) = e^{V'_{bias} - \psi'(x)} \tag{6.22}$$

and where

$$C'_{dop}(x) = N_D'^+ - N_A'^- \tag{6.23}$$

TABLE 6.3

General scaling factors.

Description	Symbol	Definition	Value	Reference
thermal voltage	V_T	$\frac{k_B T}{q}$	$0.0259V$	[7]
length scale	x_0	N/A	$L_{D,int}$	[8]
concentration	C_0	N/A	n_{int}	[8]

In obtaining the above relations we have assumed that

$$x_0 = L_{D,int} \tag{6.24}$$

where $L_{D,int}$ is known as the intrinsic Debye length [1]. Fundamental Debye length is defined as

$$L_D = \sqrt{\frac{\epsilon_0 k_B T}{q^2 C_0}} \tag{6.25}$$

Debye length is the shielding distance within the electron's plasma, where the electric field created by the perturbing charge falls off by a factor $1/e$ [9]. The mesh size used in the simulations must therefore be smaller than the Debye length in order to resolve the charge variations in space. Typical value of L_D is around $120 Angstroms$ for doping level $10^{16} cm^{-3}$. In the simulations here we choose $C_0 = n_{int}$ and therefore $L_{D,int}$ is

$$L_{D,int} = \sqrt{\frac{\epsilon_0 k_B T}{q^2 n_{int}}} \tag{6.26}$$

One can also introduce Debye lengths associated with doping as

$$L_{Dn} = \sqrt{\frac{\epsilon_0 k_B T}{q^2 N_D}} \tag{6.27}$$

and

$$L_{Dp} = \sqrt{\frac{\epsilon_0 k_B T}{q^2 N_A}} \tag{6.28}$$

In the following we *drop primes* which is a common practice.

6.4   Boundary Conditions and Trial Values

In order to solve device equations, proper boundary conditions must be formulated. Generally, boundary conditions involve transport across metallic electrodes and across heterointerfaces.

Here we concentrate on boundary conditions for electrostatic potential ψ. Complete boundary conditions involving n and p will be discussed later.

6.4.1   Boundary conditions for electrostatic potential

In Fig. 6.5 we show distribution of electrostatic potential ψ and quasi-Fermi potentials ϕ_n and ϕ_p across p-n junction [3]. Far away from the junction potential becomes constant, i.e. $d\psi/dx = 0$ away from junction. To determine

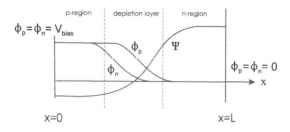

FIGURE 6.5
Distribution of potentials.

boundary conditions for ψ we assume ohmic contacts located at $x = 0$ and $x = L$. At contacts one has charge neutrality, i.e.

$$p - n + C_{dop} = 0 \qquad (6.29)$$

Also, at contacts one observes:
  Left contact $x = 0$: Here one has

$$\phi_p(0) = \phi_n(0) = V_{bias} \qquad (6.30)$$

  Right contact $x = L$: Here one has

$$\phi_p(L) = \phi_n(L) = 0 \qquad (6.31)$$

After scaling of the above relations and using charge neutrality condition (6.29) one obtains (again, dropping primes)

$$e^{\phi_p - \psi} - e^{\psi - \phi_n} + C_{dop} = 0 \qquad (6.32)$$

From above one has:
  Left contact $x = 0$:

$$e^{V_{bias} - \psi(0)} - e^{\psi(0) - V_{bias}} + C_{dop}(0) = 0$$

Solving the above, one finds

$$\psi(0) = V_{bias} + \ln\left\{ \frac{1}{2} C_{dop}(0) \left[1 + \sqrt{1 + \frac{4}{C_{dop}^2(0)}} \right] \right\} \qquad (6.33)$$

  Right contact $x = L$:

$$e^{-\psi(L)} - e^{\psi(L)} + C_{dop}(L) = 0$$

Again, the solution is

$$\psi(L) = \ln\left\{ \frac{1}{2} C_{dop}(L) \left[1 + \sqrt{1 + \frac{4}{C_{dop}^2(L)}} \right] \right\} \qquad (6.34)$$

6.4.2 Initial (trial) values for potential

Trial values of the potential are determined by extending, say relation (6.34) for contact for the whole device, as [3]

$$\psi(x_i) = \ln\left\{\frac{1}{2}\,C_{dop}(x_i)\left[1 + \sqrt{1 + \frac{4}{C_{dop}^2(x_i)}}\right]\right\} \tag{6.35}$$

where $C_{dop}(x_i)$ is the (scaled) net doping at location x_i. In the above, ψ is also scaled.

6.5 Poisson Equation for Homojunction

We will describe procedure of numerical solution of Eq.(6.20) assuming uniform value of ε_r across the sample. In such case (scaled) Poisson equation becomes (dropping primes)

$$-\varepsilon_r \frac{d^2\psi}{dx^2} = (p(x) - n(x) + C_{dop}(x)) \tag{6.36}$$

In this section we will discuss two possible methods related to different introduction of boundary conditions. We also summarize two possible ways to linearize Poisson equation.

6.5.1 Method on: Contacts outside

In this method we perform *discretization first and next we linearize* resulting equation. We consider equilibrium situation. We also place contacts outside of the computational domain, as shown in Fig. 6.6. Combining Eqs. (6.36), (6.15), and (6.16) (in scaled form), one obtains

$$-\varepsilon_r \frac{d^2\psi}{dx^2} = e^{V_{bias}-\psi(x)} - e^{\psi(x)} + C_{dop}(x) \tag{6.37}$$

At this point we introduce 1D mesh, as shown in Fig. 6.6.

Internal points of the device are numbered by i, $1 \le i \le N$. Points $i = 0$ and $i = N + 1$ represent contacts.

LHS of the above is discretized in a standard way

$$\frac{d^2\psi}{dx^2} = \frac{\psi_{i+1} - 2\psi_i + \psi_{i-1}}{\Delta x^2} \tag{6.38}$$

where $\Delta x = x_{i+1} - x_i$ and Poisson equation becomes

$$-\varepsilon_r \frac{\psi_{i+1} - 2\psi_i + \psi_{i-1}}{\Delta x^2} = e^{V_{bias}-\psi_i} - e^{\psi_i} + C_{dop}(x_i) \tag{6.39}$$

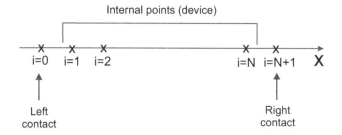

FIGURE 6.6
Uniform mesh used in method one.

Next, we perform linearization of the above equation. For that, we assume

$$\psi_i = \psi_i^0 + \delta\psi_i \qquad (6.40)$$

where $\delta\psi_i$ is a small correction. Typical exponential term at mesh point x_i takes the form

$$e^{\pm\psi_i} = e^{\pm\psi_i^0}e^{\pm\delta\psi_i} \simeq e^{\pm\psi_i^0}\left(1 \pm \delta\psi_i\right) \qquad (6.41)$$

Here ψ_i^0 represents original value of the potential (in the first step it is the trial value; in the following steps it is the value before improvement) and $\delta\psi_i$ represents improvement. Substituting the above expansion into Poisson equation and moving correction terms to the left one obtains

$$\delta\psi_{i+1} - \delta\psi_i\left(2 - e^{-\psi_i^0} - e^{\psi_i^0}\right) + \delta\psi_{i-1}$$
$$= \psi_{i+1}^0 + 2\psi_i^0 + \psi_{i-1}^0 + \frac{\Delta x^2}{\varepsilon_r}\left[e^{V_{bias}-\psi_i^0} - e^{\psi_i^0} + C_{dop}(x_i)\right] \quad (6.42)$$

The above can be written in a condensed form as

$$\delta\psi_{i-1} + b_1(i)\delta\psi_i + \delta\psi_{i+1} = f_1(i), \text{ for } i = 1, 2, 3, ...N \qquad (6.43)$$

where (index 'i' on b_1 and f_1 refers to method one)

$$b_1(i) = -\left[2 + \frac{\Delta x^2}{\varepsilon_r}\left(e^{V_{bias}-\psi_i^0} + e^{\psi_i^0}\right)\right] \qquad (6.44)$$

$$f_1(i) = -\frac{\Delta x^2}{\varepsilon_r}\left[e^{V_{bias}-\psi_i^0} - e^{\psi_i^0} + C_{dop}(x_i)\right] - \psi_{i-1}^0 + 2\psi_i^0 - \psi_{i+1}^0 \quad (6.45)$$

The above equation for several values of i is

$$\delta\psi_0 - b_1(1)\delta\psi_1 + \delta\psi_2 = f_1(1)$$
$$\delta\psi_1 - b_1(2)\delta\psi_2 + \delta\psi_3 = f_1(2)$$
$$\vdots \qquad\qquad (6.46)$$
$$\delta\psi_{N-2} - b_1(N-1)\delta\psi_{N-1} + \delta\psi_N = f_1(N-1)$$
$$\delta\psi_{N-1} - b_1(N)\delta\psi_N + \delta\psi_{N+1} = f_1(N)$$

The corrections $\delta\psi_0$ and $\delta\psi_{N+1}$ vanish since they are determined by boundary conditions as $\psi_0 = \psi_0^0$ and $\psi_{N+1} = \psi_{N+1}^0$. Boundary conditions also apply to function $f_1(i)$. The expressions close to contacts are

$$f_1(1) = -\frac{\Delta x^2}{\varepsilon_r}\left[e^{V_{bias}-\psi_1^0} - e^{\psi_1^0} + C_{dop}(x_1)\right] - \psi_0^0 + 2\psi_1^0 - \psi_2^0 \qquad (6.47)$$

and

$$f_1(N) = -\frac{\Delta x^2}{\varepsilon_r}\left[e^{V_{bias}-\psi_N^0} - e^{\psi_N^0} + C_{dop}(x_N)\right] - \psi_{N-1}^0 + 2\psi_N^0 - \psi_{N+1}^0 \qquad (6.48)$$

In the above, ψ_0^0 and ψ_{N+1}^0 are fixed by boundary conditions (values at contacts). Their values are

$$\psi_0^0 = V_{bias} + \ln\left[\frac{1}{2}C_{dop}(x_0) + \sqrt{1 + \frac{1}{4}C_{dop}^2(x_0)}\right] \qquad (6.49)$$

$$\psi_{N+1}^0 = \ln\left[\frac{1}{2}C_{dop}(x_{N+1}) + \sqrt{1 + \frac{1}{4}C_{dop}^2(x_{N+1})}\right] \qquad (6.50)$$

We assume

$$C_{dop}(x_0) = C_{dop}(x_1) \qquad (6.51)$$

and

$$C_{dop}(x_{N+1}) = C_{dop}(x_N) \qquad (6.52)$$

Trial values are selected as

$$\psi_i^{trial} = \ln\left\{\frac{1}{2}C_{dop}(x_i)\left[1 + \sqrt{1 + \frac{4}{C_{dop}^2(x_i)}}\right]\right\} \qquad (6.53)$$

The above equations can be expressed in the matrix form as

$$\begin{bmatrix} b_1(2) & 1 & & & \\ 1 & b_1(3) & 1 & & \\ & 1 & b_1(4) & 1 & \\ & & \cdots & & \\ & & & \cdots & \\ & & & 1 & b_1(N-1) \end{bmatrix}\begin{bmatrix} \delta\psi_2 \\ \delta\psi_3 \\ . \\ . \\ . \\ \delta\psi_{N-1} \end{bmatrix} = \begin{bmatrix} f_1(2) \\ f_1(3) \\ . \\ . \\ . \\ f_1(N-1) \end{bmatrix}$$
$$(6.54)$$

The above matrices written symbolically are

$$\overleftrightarrow{A}\,\overrightarrow{\delta\psi} = \overrightarrow{f} \qquad (6.55)$$

The implementation of the above system and the analysis is provided is provided later.

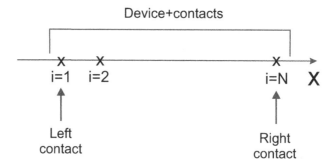

FIGURE 6.7
Uniform mesh used in method two.

6.5.2   Method two: Contacts inside

In this case contacts are defined as $i = 1$ and $i = N$, i.e. they are within computational domain, as shown in Fig. 6.7. In this method, we linearize Poisson equation first and then discretize it. For this case we also linearize Poisson equation differently [1]. We consider equilibrium situation. Poisson equation is (neglecting external bias voltage)

$$
\begin{aligned}
-\varepsilon_r \frac{d^2\psi}{dx^2} &= p(x) - n(x) + C_{dop}(x) \\
&= e^{-\psi(x)} - e^{\psi(x)} + C_{dop}(x)
\end{aligned}
\tag{6.56}
$$

Evaluate potential ψ in the iteration steps. Assume that $n+1$ refers to the values at the next step, and n refers to the values at the previous step. Poisson equation thus can be written as

$$
-\varepsilon_r \frac{d^2\psi^{(n+1)}}{dx^2} = e^{-\psi^{(n+1)}} - e^{\psi^{(n+1)}} + C_{dop}
\tag{6.57}
$$

Assume as before

$$
\psi^{(n+1)} = \psi^{(n)} + \delta
\tag{6.58}
$$

where δ is a small correction. Typical exponential term at the RHS at mesh point x_i takes the form

$$
e^{\pm\psi^{(n)}} = e^{\pm\psi^{(n)}} e^{\pm\delta} \simeq e^{\pm\psi^{(n)}} (1 \pm \delta)
\tag{6.59}
$$

With the above assumptions, Poisson eq becomes

$$
\begin{aligned}
&-\varepsilon_r \frac{d^2\psi^{(n+1)}}{dx^2} - \psi^{(n+1)} \left(e^{-\psi^{(n)}} + e^{\psi^{(n)}} \right) \\
&= \left[e^{-\psi^{(n)}} - e^{\psi^{(n)}} + C_{dop} \right] - \delta \left(e^{-\psi^{(n)}} + e^{\psi^{(n)}} \right) \\
&= \left[e^{-\psi^{(n)}(x)} - e^{\psi^{(n)}(x)} + C_{dop}(x) \right] + \psi^{(n)} \left(e^{-\psi^{(n)}} + e^{\psi^{(n)}} \right)
\end{aligned}
\tag{6.60}
$$

where we have used that $\delta = \psi^{(n+1)} - \psi^{(n)}$. The above can also be written in terms of densities which are evaluated at the previous step as

$$-\varepsilon_r \frac{d^2 \psi^{(n+1)}}{dx^2} - \psi^{(n+1)} \left(p^{(n)} + n^{(n)}\right) = \left(p^{(n)} - n^{(n)} + C_{dop}\right) - \psi^{(n)} \left(p^{(n)} + n^{(n)}\right)$$
(6.61)

The above equation describes iteration process (i.e. new value of potential at step $n+1$ in relation to a previous value at step n). It can be discretized and one obtains

$$\psi_{i+1}^{(n+1)} - \left[2 + \frac{\Delta x^2}{\varepsilon_r} \left(p_i^{(n)} + n_i^{(n)}\right)\right] \psi_i^{(n+1)} + \psi_{i-1}^{(n+1)}$$

$$= -\frac{\Delta x^2}{\varepsilon_r} \left[\left(p_i^{(n)} - n_i^{(n)} + C_{dop}(x_i)\right) + \psi_i^{(n)} \left(p_i^{(n)} + n_i^{(n)}\right)\right] \quad (6.62)$$

The above in a condensed form is

$$a_2(i)\psi_{i+1}^{(n+1)} + b_2(i)\psi_i^{(n+1)} + c_2(i)\psi_{i-1}^{(n+1)} = f_2^{(n)}(i) \qquad (6.63)$$

where

$$a_2(i) = 1, \quad b_2(i) = -\left[2 + \frac{\Delta x^2}{\varepsilon_r} \left(p_i^{(n)} + n_i^{(n)}\right)\right], \quad c_2(i) = 1 \qquad (6.64)$$

and

$$f_2^{(n)}(i) = -\frac{\Delta x^2}{\varepsilon_r} \left[\left(p_i^{(n)} - n_i^{(n)} + C_i\right) + \psi_i^{(n)} \left(p_i^{(n)} + n_i^{(n)}\right)\right] \qquad (6.65)$$

Close to contacts, the above equation produces

$$a_2(1)\psi_0 + b_2(1)\psi_1 + c_2(1)\psi_2 = f_2^{(n)}(1)$$
$$a_2(2)\psi_1 - b_2(2)\psi_2 + c_2(2)\psi_3 = f_2^{(n)}(2)$$
$$\vdots \qquad\qquad (6.66)$$
$$a_2(N-1)\psi_{N-2} - b_2(N-1)\psi_{N-1} + c_2(N-1)\psi_N = f_2^{(n)}(N-1)$$
$$a_2(N)\psi_{N-1} - b_2(N)\psi_N + c_2(N)\psi_{N+1} = f_2^{(n)}(N)$$

Observe that $i = 1$ corresponds to left contact and that the value of potential there, ψ_1 is established by boundary conditions. One is therefore left with the following boundary conditions

$$a_2(1) = c_2(1) = 0, \quad \text{and} \quad b_2(1) = 1 \qquad (6.67)$$

The above establishes the value of $f_2(1)$ as

$$f_2^{(n)}(1) = \psi_1 \qquad (6.68)$$

In the same way boundary conditions are establish at right contact, i.e. for $i = N$

$$a_2(N) = c_2(N) = 0, \quad \text{and} \quad b_2(N) = 1 \qquad (6.69)$$

and

$$f_2^{(n)}(N) = \psi_N \qquad (6.70)$$

6.6   Poisson Equation for Non-Uniform Systems

We use interchangeably (where convenient) the notation x_i and i. Poisson equation (6.20) in scaled form is written as (dropping primes)

$$-\frac{d}{dx}J_v = p - n + C_{dop} \tag{6.71}$$

where the net doping is

$$C_{dop} = (N_D^+ - N_A^-) \tag{6.72}$$

and Poisson flux J_v is defined as

$$J_v = \varepsilon_r \frac{d}{dx}\psi \tag{6.73}$$

6.6.1   Linearization

Linearization is performed like in Method two, following [1]. We express potential in terms of old value as

$$\psi^{(n+1)} = \psi^{(n)} + \delta \tag{6.74}$$

where δ is a small correction and index n signifies that the value was determined at the previous step. Terms related to densities are evaluated as

$$e^{\pm\psi^{(n+1)}} = e^{\pm\psi^{(n)}}e^{\pm\delta} = e^{\pm\psi^{(n)}}\left(1 \pm \delta\right) \tag{6.75}$$

The LHS which contains only new (updated) value of the potential is

$$LHS = -\frac{d}{dx}\varepsilon_r \frac{d}{dx}\psi^{(n+1)} \tag{6.76}$$

The RHS is linearized as follows

$$\begin{aligned}
RHS &= p - n + C_{dop} = e^{-\psi^{(n+1)}} - e^{\psi^{(n+1)}} + C_{dop}\\
&= e^{-\psi^{(n)}} - e^{\psi^{(n)}} + C_{dop} - \delta\left(e^{-\psi^{(n)}} + e^{\psi^{(n)}}\right)\\
&= e^{-\psi^{(n)}} - e^{\psi^{(n)}} + C_{dop} - \psi^{(n+1)}\left(e^{-\psi^{(n)}} + e^{\psi^{(n)}}\right)\\
&+ \psi^{(n)}\left(e^{-\psi^{(n)}} + e^{\psi^{(n)}}\right)
\end{aligned} \tag{6.77}$$

Combining the above results allows to write (linearized) Poisson equation as

$$\begin{aligned}
-\frac{d}{dx}J_v^{(n+1)} + \psi^{(n+1)}\left(e^{-\psi^{(n)}} + e^{\psi^{(n)}}\right) &= e^{-\psi^{(n)}} - e^{\psi^{(n)}}\\
+ C_{dop} + \psi^{(n)}\left(e^{-\psi^{(n)}} + e^{\psi^{(n)}}\right)&
\end{aligned} \tag{6.78}$$

Observe that at equilibrium $p^{(n)} = e^{-\psi^{(n)}}$ and $n^{(n)} = e^{\psi^{(n)}}$ and therefore Poisson equation can be expressed as

$$-\frac{d}{dx}J_v^{(n+1)} + \psi^{(n+1)}\left(p^{(n)} + n^{(n)}\right) = p^{(n)} - n^{(n)} + C_{dop} + \psi^{(n)}\left(p^{(n)} + n^{(n)}\right)$$
(6.79)

6.6.2 Discretization

The derivative of Poisson flux $J_v^{(n+1)}$ can be discretized as, see Figure 6.8

$$\frac{d}{dx}J_v^{(n+1)} = \frac{J_v^{(n+1)}(i + \frac{1}{2}) - J_v^{(n+1)}(i - \frac{1}{2})}{\Delta x_i}$$
(6.80)

where

$$\begin{aligned}
\Delta x_i &= x_{i+1/2} - x_{i-1/2} \\
&= \frac{1}{2}(x_{i+1} + x_i) - \frac{1}{2}(x_i + x_{i-1}) \\
&= \frac{1}{2}(x_{i+1} - x_{i-1})
\end{aligned}$$
(6.81)

We account for the nonuniformity of Poisson flow due to nonuniformity of dielectric constant appearing in Eq. (6.73). We therefore discretize Poisson flux as (we have dropped index r on ε)

$$J_v^{(n+1)}\left(i + \frac{1}{2}\right) = \frac{\varepsilon_{i+1}\psi_{i+1}^{(n+1)} - \varepsilon_i\psi_i^{(n+1)}}{x_{i+1} - x_i}$$
(6.82)

and

$$J_v^{(n+1)}\left(i - \frac{1}{2}\right) = \frac{\varepsilon_i\psi_i^{(n+1)} - \varepsilon_{i-1}\psi_{i-1}^{(n+1)}}{x_i - x_{i-1}}$$
(6.83)

Details of mesh across heterointerface are shown in Fig. 6.9. Values of dielectric constant ε_r (index r dropped) are shown at three mesh points which belong to two different layers, here labeled as layer "m" and layer "$m + 1$". At each of those mesh points there are also well-defined values of electrostatic potential ψ.

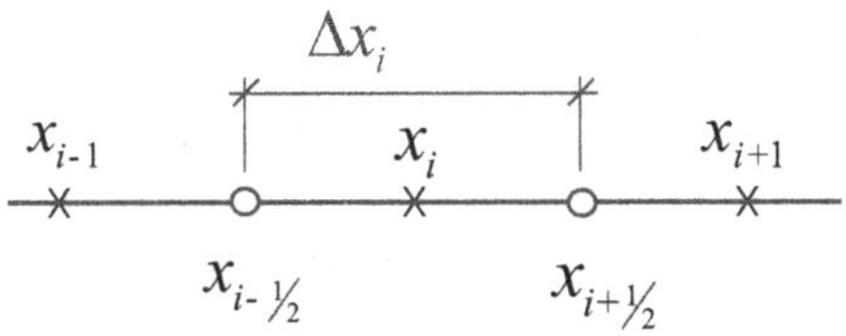

FIGURE 6.8
Notation for intermediate mesh points.

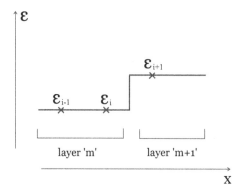

FIGURE 6.9
Mesh details across heterointerface.

Combining the above results, the discretized Poisson equation becomes

$$\frac{\varepsilon_{i-1}}{x_i - x_{i-1}}\psi_{i-1}^{(n+1)} - \left[\left(\frac{\varepsilon_i}{x_{i+1} - x_i} + \frac{\varepsilon_i}{x_i - x_{i-1}}\right)\right.$$
$$\left. + \Delta x_i \left(e^{-\psi^{(n)}} + e^{\psi^{(n)}}\right)\right]\psi_i^{(n+1)}$$
$$+ \frac{\varepsilon_{i+1}}{x_{i+1} - x_i}\psi_{i+1}^{(n+1)}$$
$$= -\Delta x_i \left[e^{-\psi_i^{(n)}} - e^{\psi_i^{(n)}} + C_{dop}(x_i) + \psi_i^{(n)}\left(e^{-\psi_i^{(n)}} + e^{\psi_i^{(n)}}\right)\right] \quad (6.84)$$

Eq. (6.84) is finally expressed in a condensed form as

$$a(i)\psi_{i-1} + b(i)\psi_i + c(i)\psi_{i+1} = f(i) \quad (6.85)$$

where the coefficients are defined as

$$a(i) = \frac{\varepsilon_{i-1}}{x_i - x_{i-1}} \quad (6.86)$$

$$b(i) = -\left[\left(\frac{\varepsilon_i}{x_{i+1} - x_i} + \frac{\varepsilon_i}{x_i - x_{i-1}}\right) + \Delta x_i \left(e^{-\psi^{(n)}} + e^{\psi^{(n)}}\right)\right] \quad (6.87)$$

$$c(i) = \frac{\varepsilon_{i+1}}{x_{i+1} - x_i} \quad (6.88)$$

$$f(i) = -\Delta x_i \left[e^{-\psi_i^{(n)}} - e^{\psi_i^{(n)}} + C_{dop}(x_i) + \psi_i^{(n)}\left(e^{-\psi_i^{(n)}} + e^{\psi_i^{(n)}}\right)\right] \quad (6.89)$$

Eq.(6.85) is supplemented by the boundary conditions for potential $\psi(x)$, see next section.

6.6.3 Boundary conditions for potential

We allow here also for an external bias voltage V_{bias}. In such case, boundary conditions were established previously (in a scaled form), see Eqs. (6.33) and (6.34). We bring them here again for convenience

$$\psi(0) = V_{bias} + \ln\left\{\frac{1}{2}C_{dop}(0)\left[1 + \sqrt{1 + \frac{4}{C_{dop}^2(0)}}\right]\right\} \qquad (6.90)$$

$$\psi(N) = \ln\left\{\frac{1}{2}C_{dop}(N)\left[1 + \sqrt{1 + \frac{4}{C_{dop}^2(N)}}\right]\right\} \qquad (6.91)$$

6.6.4 Initial conditions for potential

Initial conditions were investigated before. Here, we summarize for completeness. Initial potential is determined as

$$\psi(x_i) = \ln\left\{\frac{1}{2}C_{dop}(x_i)\left[1 + \sqrt{1 + \frac{4}{C_{dop}^2(x_i)}}\right]\right\} \qquad (6.92)$$

6.7 Applications of Poisson Equation to Analyze p-n Diode

6.7.1 General

In this chapter we consider only homo-diode, that is uniform p-n junction. It is modeled using only Poisson equation and Boltzmann-type expressions for densities of electrons and holes. This type of analysis (and similar) has been reported by several groups [2, 10, 11].

Flowchart for solving one-dimensional Poisson equation is shown in Fig. 6.10. In the following sections we discuss specific cases and compare with the published results. List of Matlab files created for Chapter 6 and a short description of each function is provided in Table 6.4.

6.7.2 Analysis of convergence

In this section, based on the previously developed formalism we demonstrate numerical solution of Poisson's equation in one dimension. Boundary conditions are not introduced here. In such an approach charge density is only function of potential. We compare our results with those of Shapiro [2]. Following Shapiro [2] we selected intrinsic concentration to be $n_{int} = 1.45 \times 10^{10} cm^{-3}$. The dopant concentrations are $N_A^- = 10^{16} cm^{-3}$ on

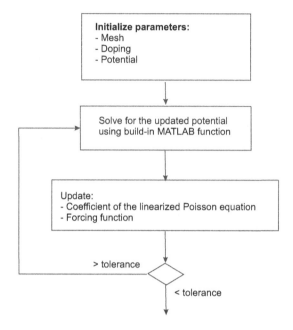

FIGURE 6.10
Flowchart for 1D solver of Poisson equation.

the p-side and $N_D^+ = 10^{16} cm^{-3}$ on the n-side. We use Matlab build-in functions to solve algebraic system of equations. Potential for several iterations for zero bias is shown in Fig. 6.11.

Note: mesh here includes contacts.

TABLE 6.4
List of Matlab functions for Chapter 6.

Listing	Function	Description
Simple Poisson		
6.A.1	*simple_poisson.m*	Solves simple (1D) Poisson equation
Convergence		
6.A.2.1	*poisson_iter.m*	Shapiro program
6.A.2.2	*param_simple.m*	Parameters for Shapiro
Linear Poisson		
6.A.3.1	*poisson_linear_zero.m*	Solves Poisson eq. for linear doping
6.A.3.2	*param_simple.m*	Parameters for linear doping Same file as for Shapiro

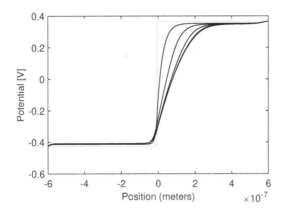

FIGURE 6.11
Potential for several iterations for zero bias. Dashed line represents an initial guess (compare results with [2]).

6.7.3 Homo-junction with linear doping

Linear doping is assumed to be

$$C_{dop}(x) = m \cdot x \qquad (6.93)$$

where x is a position within the junction. We consider case without external bias voltage. We introduced here boundary conditions. The results for electrostatic potential, along with initial guess (dotted line) are shown in Fig. 6.12.

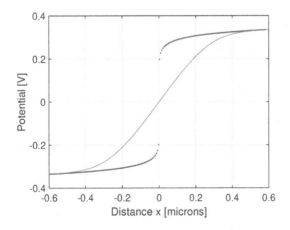

FIGURE 6.12
Potential for homo-junction with linear doping for zero bias. Dashed line represents an initial guess.

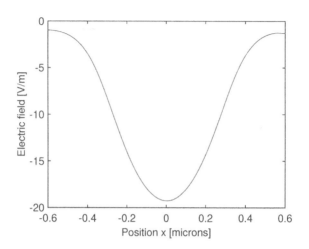

FIGURE 6.13
Electric field for homo-junction with linear doping for zero bias.

In Fig. 6.13 we show the profile of electric field for the same structure. Our results compares well with Jiao et al [11].

The results for electric field are not extended to regions where field vanishes,i.e. where

$$d\psi/dx(x = \pm a) = 0 \qquad (6.94)$$

Jiao et al [11] determined values of a for various values of external potential.

Appendix 6A: MATLAB Listings

Listing 6.A.1
Program solves Poisson equation using a second-order central difference scheme and compares it with an exact solution.

```
% File name: poisson_simple.m
% Solves one-dimensional Poisson equation
% u''=-x^2
% with 0<x<1
% u(0)=0
% u(1)=0
% using a second-order central difference scheme
% and compares with the exact solution.
% Exact solution u_exact is
% u_exact = (-1/12)*x^4 + (1/12)*x
```

```
%
clear all
N_mesh = 10;        % number of mesh points
a = 0.0;            % left coordinate
b = 1.0;            % right coordinate
u_a = 0;            % potential value at left coordinate
u_b = 0;            % potential value at right coordinate
%
h = (b - a)/(N_mesh+1);             % mesh size
%
% x(0)= a, is the value at the left boundary
% x(N_mesh+1) = b, is the value at right boundary
%
for k = 1:N_mesh
    x(k) = a + k*h;
end

% Creation of matrix A
for i=1:N_mesh
    A(i,i) = -2.0;          % diagonal elements
end
for i=1:N_mesh-1
    A(i,i+1) = 1.0;         % above diagonal
    A(i+1,i) = 1.0;         % below diagonal
end
%
% Creation of RHS vector
for k=1:N_mesh
    g(k)=-h^2*x(k)^2;
end
% Solution
u = g/A;        % solution obtained using Matlab functions
% For plotting purposes, we also add boundary points
x_plot = [a,x,b];
u_plot = [u_a,u,u_b];
%
% Exact solution u_exact is
x_exact = linspace(a,b,100);
u_exact = (-1/12)*x_exact.^4 + (1/12)*x_exact;
plot(x_plot,u_plot,x_exact,u_exact,'- .')
xlabel('x value','Fontsize',16);
ylabel('Solution','Fontsize',16);
set(gca,'FontSize',14);         % size of tick marks on both axis
pause
close all
```

Listing 6.A.2.1
Shapiro program for test of convergence.

```
% File name: homo_iter.m
% Solves one-dimensional Poisson equation for p-n junction
% (linear doping)
% without bias voltage.
% Illustrates convergence. Comparison with Shapiro (1995)
%
clear all
param_simple;  % input parameters for computation
%
% Generate mesh; values of mesh points are scaled
 x = zeros(1,N_mesh);
 x(1) = a;
for k = 2:N_mesh
    x(k) = a + (k-1)*h;
end
x(N_mesh)= b;
%--------- define doping
for i = 1:N_mesh
    if(i <= N_mesh/2)
        dop(i) = - N_A/n_int;
    elseif(i > N_mesh/2)
        dop(i) = N_D/n_int;
    end
end
%------------------------------
psi_trial = zeros(1,N_mesh); % Initialize array to hold trial potential
delta_psi = zeros(1,N_mesh); % Initialize array to hold corrections
relative_error = 1.0;         % Initial value of the relative error
%------------------------------
% Assign trial values at internal mesh points
% Values at contacts are established by boundary conditions and
% will not
% change during computations
for i = 1:N_mesh
    dd = dop(i);
    psi_trial(i) = log(dd + sqrt(1.0 + dd^2));
end
plot(x_zero*x,V_T*psi_trial,'r:');  % plot trial potential
pause
hold on
psi = psi_trial;
sigma = zeros(1,N_mesh-2);        % Initialize array to hold potential
%------------------------------
% Creation of matrix A
A = zeros(N_mesh-2,N_mesh-2);   % Initialize array to hold matrix A
count = 0;
```

```
while (relative_error >= error_iter)


for i = 1:N_mesh-2
  sigma(i)=-2.0-(h^2)/(eps_r*lambda_sq)*
  ( exp(-psi(i+1))+exp(psi(i+1)) );
end


for i=1:N_mesh-2
   A(i,i) = sigma(i);          % diagonal elements
end
for i=1:N_mesh-3
   A(i,i+1) = 1.0;             % above diagonal
end
for i=1:N_mesh-3
   A(i+1,i) = 1.0;             % below diagonal
end
%-----------------------------
% Creation of rhs vector
f = zeros(1,N_mesh-2);         % Initialize array to hold rhs vector
for i=2:N_mesh-1
   aa = dop(i)+exp(-psi(i)) - exp(psi(i));
   f(i-1)=-psi(i-1)+2.0*psi(i)-psi(i+1)-(h^2*aa)/(eps_r*lambda_sq);
end
delta_psi(1,2:N_mesh-1) = f/A;  % Solve for correction potential
% The solution is obtained and put in a vector without end mesh points.
% End points are held by boundary conditions
psi = psi + delta_psi;          % New value of potential
relative_error = max(abs(delta_psi(1,2:N_mesh-1)./psi(1,2:N_mesh-1)));
count = count +1;
plot(x_zero*x,V_T*psi,'k-');    % Plot all results of iterations
end          % End of while loop
%
%plot(x_zero*x,V_T*psi,'k-');    % Plot only final iteration
%plot(x_zero*x,V_T*psi);
xlabel('Position (meters)','Fontsize',14);
ylabel('Potential [V]','Fontsize',14);
set(gca,'FontSize',14);         % size of tick marks on both axis
pause
close all
```

Listing 6.A.2.2
Parameters for Shapiro (convergence) test.

```
% File name: param_simple.m
% Purpose:
% Contains scaled parameters for p-n junction
% Description is based on Poisson equation only
%
```

```
% Definitions of basic physical constants
epsilon_zero = 8.8541878d-12; % permittivity of free space; unit [F/m]
q = 1.6021892d-19;            % electron's charge; unit [Coulombs]
k_B = 1.380662d-23;           % Boltzman constant; unit [J/K]
T = 300;                      % absolute temperature; unit [K]
%---------------------------
% Material's parameters
n_int = 1.4d16;               % intrinsic concentration; unit [m^-3]
eps_r = 11.7;                 % relative dielectric constant for Si
epsil = epsilon_zero*eps_r;
%---------------------------
% Doping parameters
m = 1.0d28;                   % parameter for linear doping; unit [m^-4]
N_A = 10d22;                  % doping for p-region; unit [m^-3]
N_D = 10d21;                  % doping for n-region; unit [m^-3]
%---------------------------
% Scaling factors
x_zero = 0.43d-6;             % scaling of distance; unit [m]
V_T = k_B*T/q;
lambda_sq = V_T*epsilon_zero/(x_zero^2*q*n_int); % parameter
m = m*x_zero/n_int;           % scaling of m (doping parameter)
%---------------------------
% Definition of structure and numerical parameters
a = -0.6d-6;                  % left coordinate of junction; unit [m]
b = 0.6d-6;                   % right coordinate of junction; unit [m]
N_mesh = 200;                 % number of mesh points
%
h = (b - a)/(N_mesh+1);       % mesh size
%
h = h/x_zero;                 % scaled mesh size
a = a/x_zero;                 % scaled left coordinate
b = b/x_zero;                 % scaled right coordinate
%
error_iter = 0.000001;        % iteration error
```

Listing 6.A.3.1
Program solves Poisson equation for p-n junction with linear doping at zero external voltage.

```
% File name:    poisson_linear_zero.m
% Solves one-dimensional Poisson equation for p-n junction
% without bias voltage.
% Linear doping is assumed
% All variables are scaled
%
clear all
close all
tic
```

```
param_simple;                    % input parameters for computation
%
% Generate mesh
x = zeros(1,N_mesh);
x(1) = a;
for k = 2:N_mesh-1
    x(k) = a + (k-1)*h;
end
x(N_mesh)= b;
%----------------------------
psi_trial = zeros(1,N_mesh); % Initialize array to hold trial potential
delta_psi = zeros(1,N_mesh); % Initialize array to hold corrections
%----------------------------
% Assign trial values at internal mesh points
% Values at contacts are established by boundary conditions and
% will not
% change during computations
for i = 1:N_mesh
    dd(i) = m*x(i)/2;  % linear doping
    psi_trial(i) = log(dd(i) + sqrt(1.0 + dd(i)*dd(i)));
end

plot(x_zero*x*1.0d6,V_T*psi_trial,'.');  % plots trial potential
pause
hold on
psi = psi_trial;
sigma = zeros(1,N_mesh);      % Initialize array to hold potential
A = zeros(N_mesh,N_mesh);     % Initialize array to hold matrix A

% Establish boundary conditions
psi_L = psi_trial(1);          % left contact
psi_R = psi_trial(N_mesh);     % right contact

a_L = m*x(1)+exp(-psi(1)) - exp(psi(1));
a_R = m*x(N_mesh)+exp(-psi(N_mesh)) - exp(psi(N_mesh));

f(1)=-psi_L+2.0*psi(1)-psi(2)-(h^2*a_L)/(eps_r*lambda_sq);
f(N_mesh) = -psi(N_mesh-1)+2.0*psi(N_mesh)-psi_R-
(h^2*a_R)/(eps_r*lambda_sq);

count = 0;
relative_error = 1.0;          % Initial value of the relative error
while (relative_error >= error_iter)
    for i = 1:N_mesh
        sigma(i)=2.0+(h^2)/(eps_r*lambda_sq)*( exp(-psi(i))
        +exp(psi(i)) );
    end
%===============================================
% Creation of matrix A
```

```
    for i=1:N_mesh
        A(i,i) = -sigma(i);              % diagonal elements
    end
    for i=1:N_mesh-1
        A(i,i+1) = 1.0;                  % above diagonal
    end
    for i=1:N_mesh-1
        A(i+1,i) = 1.0;                  % below diagonal
    end
% Creation of rhs vector
    f = zeros(1,N_mesh);                 % Initialize array to hold rhs vector
    for i=2:N_mesh-1
        aa = m*x(i)+exp(-psi(i)) - exp(psi(i));
        f(i)=-psi(i-1)+2.0*psi(i)-psi(i+1)-(h^2*aa)/(eps_r*lambda_sq);
    end

    delta_psi(1:N_mesh) = f/A;  % Solve for correction potential
    psi = psi + delta_psi;               % New value of potential
    relative_error = max(abs(delta_psi(1,1:N_mesh)./psi(1,1:N_mesh)));
    count = count +1;
end

toc

plot(x_zero*x*1.0d6,V_T*psi);            % Plot final result of iterations
xlabel('Distance x [microns]','Fontsize',16);
ylabel('Potential [V]','Fontsize',16);
set(gca,'FontSize',14);                  % size of tick marks on both axis
grid on;
pause
close all
% Determines electric field
for i=2:N_mesh-1
    electric_field(i)=-(psi(i)-psi(i-1))/h;
end
electric_field(1)=electric_field(2);
electric_field(N_mesh)=electric_field(N_mesh-1);
plot(x_zero*x*1.0d6,electric_field);
xlabel('Position x [microns]','Fontsize',14);
ylabel('Electric field [V/m]','Fontsize',14);
set(gca,'FontSize',14);                  % size of tick marks on both axis
pause
close all
```

Bibliography

[1] Vasileska D., Goodnick S.M., and Klimeck G. *Computational Electronics.* CRC Press, Boca Raton, 2010.

[2] F.R. Shapiro. The numerical solution of Poisson equation in a pn diode using a spreadsheet. *IEEE Transactions on Education*, 38:380–384, 1995.

[3] M. Kurata. *Numerical Analysis for Semiconductor Devices.* Lexington-Books, New York, 1982.

[4] I.D. Mayergoyz. Solution of the nonlinear poisson equation of semiconductor device theory. *J. Appl. Phys.*, 59:195–199, 1986.

[5] C.M. Krowne. Semiconductor heterostructure nonlinear Poisson equation. *J. Appl. Phys.*, 65:1602–1614, 1989.

[6] A. Jozwikowska. Numerical solution of the nonlinear Poisson equation for semiconductor devices by application of a diffusion-equation finite difference scheme. *J. Appl. Phys.*, 104:063715, 2008.

[7] S. Selberherr. *Analysis and Simulation of Semiconductor Devices.* Springer, Wien, New York, 1984.

[8] S.J. Polak, C. Den Heijer, and W.H.A. Schilders. Semiconductor device modelling from the numerical point of view. *International Journal for Numerical Methods in Engineering*, 24:763–838, 1987.

[9] R.A. Pierret. *Semiconductor Device Fundamentals.* Addison-Wesley, Reading, Massachusetts, 1996.

[10] R.A. Jabr, M. Hamad, and Y.M. Mohanna. Newton-Raphson solution of Poisson's equation in a pn diode. *The International Journal of Electrical Engineering and Education*, 44:22–33, 2007.

[11] Y.C. Jiao, C. Dang, and Y. Hao. The solution of the one-dimensional nonlinear Poisson's equations by the decomposition method. *An International Journal Computers and Mathematics with Applications*, 46:1645–1656, 2003.

7

Experiments Using Poisson Equation: Homo diode

We use previously developed methods to numerically solve Poisson equation for p-n diode. As another application of Poisson equation we discuss p-n junction with *step doping* in equilibrium. In this chapter we consider homo structure and analyze it using two methods. Methods differ by way how we incorporate contacts into numerical scheme and also how we linearize and discretize Poisson equation. Hetero structure will be discussed in the next chapter.

Starting with this chapter we reorganize program and split it into files, each responsible for a particular functionality. Home diode with step doping is essentially the same program as the previous one for linear doping but here files are organized according to functionality. In the following chapters we will keep the same idea of organizing files according to functionalities.

We also created separate file which deals with contacts (boundary.m). In this way one has the flexibility to design (if needed) various types of contacts.

7.1 Method One

Method with contacts outside. Contacts are numbered as $i = 0$ (left contact) and $i = N + 1$ (right contact).

For completeness we start with a summary of the equations which will be implemented. The corrections to the potential are calculated using the following (N is the number of mesh points; contacts are located at $i = 0$ (left contact) and $i = N + 1$ (right contact))

$$\delta\psi_{i-1} + b_1(i)\delta\psi_i + \delta\psi_{i+1} = f_1(i), \qquad i = 1, 2, 3, \ldots N \tag{7.1}$$

where

$$b_1(i) = -[2 + \frac{\Delta x^2}{\varepsilon_r}\left(e^{V_{bias} - \psi_i^0} + e^{\psi_i^0}\right)] \tag{7.2}$$

$$f_1(i) = -\frac{\Delta x^2}{\varepsilon_r}\left[e^{V_{bias} - \psi_i^0} - e^{\psi_i^0} + C_{dop}(x_i)\right] - \psi_{i-1}^0 + 2\psi_i^0 - \psi_{i+1}^0 \tag{7.3}$$

DOI: 10.1201/9781003265849-7

In the above ψ_i^0 is the value at mesh point i determined at the previous iteration step. Close to contacts, expressions for the RHS are

$$f_1(1) = -\frac{\Delta x^2}{\varepsilon_r}\left[e^{V_{bias}-\psi_1^0} - e^{\psi_1^0} + C_{dop}(x_1)\right] - \psi_0^0 + 2\psi_1^0 - \psi_2^0$$

and

$$f_1(N) = -\frac{\Delta x^2}{\varepsilon_r}\left[e^{V_{bias}-\psi_N^0} - e^{\psi_N^0} + C_{dop}(x_N)\right] - \psi_{N-1}^0 + 2\psi_N^0 - \psi_{N+1}^0$$

In the above last two equations the values at contacts are

$$\psi_0^0 = V_{bias} + \ln\left\{\frac{1}{2}C_{dop}(x_0)\left[1 + \sqrt{1 + \frac{4}{C_{dop}(x_0)}}\right]\right\} \qquad (7.4)$$

$$\psi_{N+1}^0 = V_{bias} + \ln\left\{\frac{1}{2}C_{dop}(x_{N+1})\left[1 + \sqrt{1 + \frac{4}{C_{dop}(x_{N+1})}}\right]\right\} \qquad (7.5)$$

In the implementation I assumed that $C_{dop}(x_0) = C_{dop}(x_1)$ and $C_{dop}(x_{N+1}) = C_{dop}(x_N)$.

The above equations can be written as

$$\overleftrightarrow{A}\,\overrightarrow{\delta\psi} = \overrightarrow{f} \qquad (7.6)$$

Solution of such problem is obtained using build-in Matlab function. Before implementation, all variables are scaled.

7.1.1 Calculations of band edges

In this program we also determine band edges for conduction and valence bands. We use the following expressions (derived previously)

$$E_c = E_0 - \chi_e - qV(x) \qquad E_v = E_0 - \chi_e - E_g - qV(x) \qquad (7.7)$$

where E_0 is the reference energy, χ_e is the electron's affinity and E_g is the bandgap.

We use the following values for silicon: reference energy $E_0 = 5eV$, electron affinity $\chi_e = 4.05eV$, bandgap $E_g = 1.12eV$. Scaling of band edges is provided in Table 7.1.

7.1.2 Comments about mesh

When building mesh used for numerical implementation one should keep in mind that dx (mesh step) should be smaller then L_D to resolve all details; here L_D is the Debye length introduced earlier. One can also use L_{Dp} and L_{Dn} which are Debye lengths in p and n regions, respectively.

TABLE 7.1

Scaling of energies.

Description	Symbol	Scaling factor
Energy reference	E_0	$k_B T$
Bandgap	E_g	$k_B T$
Electron affinity	χ_e	$k_B T$

Created mesh should be able to resolve fine details of the device. Therefore, the size should be smaller then Debye length in the appropriate region, i.e.

$$\Delta x < L_{Debye} \tag{7.8}$$

Mesh fragment which accounts for it is provided below

```
% Definition of structure
a = 0.0;        % left coordinate of junction; unit [cm]
b = 1.0d-4;     % right coordinate of junction; unit [cm];1 micron
N_init = 201;   % initial number of mesh points
%
dx_in = (b-a)/(N_init-1);  % initial mesh size
% Compare mesh size with Debye length
temp = min(L_Dp,L_Dn);
if (dx_in > temp)
    ratio = dx_in/temp;
    N_mesh = round(N_init*ratio);
    dx = temp;
else
    N_mesh = N_init;
end
```

The above program creates very fine mesh. However, number of mesh points is extremely large. During development of the program we observed that it is not necessary to create such fine mesh. Instead, we opted for used-defined number of mesh points in the relevant regions.

7.1.3   Description of functions

Condensed list of functions is summarized in Table 7.2. Below is a more detailed description of the functions used:

- boundary.m.

  This function provides values of potential and also densities of electrons and holes (for future use) at contacts. Ohmic contacts are assumed. Contacts are numbered as $i = 0$ (left) and $i = N + 1$ (right). All variables used in this function are scaled.

TABLE 7.2

List of Matlab functions for Chapter 7. Homo junction in equilibrium as discussed by two methods.

Listing	Function	Description
Common files		
7.A.1.1	*boundary_psi.m*	Determines boundary conditions (ohmic)
7.A.1.2	*doping_step.m*	Determines step doping
7.A.1.3	*homo_poisson.m*	Driver for both methods
7.A.1.4	*initial_psi.m*	Initializes potential
7.A.1.5	*mesh_uniform.m*	Creates mesh
7.A.1.6	*physical_const.m*	Contains physical constants
7.A.1.7	*scaling_simple.m*	Contains scaling factors
Method 1		
7.A.2.1	*solution_psi_1.m*	Solves for potential
Method 2		
7.A.3.1	*solution_psi_2.m*	Solves for potential, Method 2

- doping_step.m.

  This function sets up the doping at all mesh points and scales the values with C_0. The p-n junction is assumed to be at the middle point.

- homo_poisson.

  This function implements 1D model of homogeneous p-n junction based on Poisson equation.

- initial.m.

  This function initializes potential (ψ) and densities based on the requirement of charge neutrality throughout the structure. Contacts are not included.

- mesh_uniform.m.

  This function creates 1D uniform mesh.

- physical_const.m.

  This function summarizes basic physical constants needed for simulations.

- scaling.m.

  This function keeps general and electrical scaling factors.

- solution_psi.m.

  This function determines potential (ψ) in the iteration process using Matlab functions. Values of ψ on input are for device + contacts.

7.2 Method Two

In this method (which we call Method two) contacts are defined to be inside the structure [1].

The above equation in a condensed form is

$$a_2(i)\psi_{i+1}^{(n+1)} + b_2(i)\psi_i^{(n+1)} + c_2(i)\psi_{i-1}^{(n+1)} = f_2^{(n)}(i) \tag{7.9}$$

where

$$a_2(i) = 1, \quad b_2(i) = -\left[2 + \frac{\Delta x^2}{\varepsilon_r}\left(p_i + n_i\right)\right], \quad c_2(i) = 1 \tag{7.10}$$

and

$$f_2^{(n)}(i) = -\frac{\Delta x^2}{\varepsilon_r}\left[\left(p_i^{(n)} - n_i^{(n)} + C_i\right) + \psi_i^{(n)}\left(p_i^{(n)} + n_i^{(n)}\right)\right] \tag{7.11}$$

Here (n) specifies iteration step. The following boundary conditions are applied

$$a_2(1) = c_2(1) = 0, \quad \text{and} \quad b_2(1) = 1 \tag{7.12}$$

The above establishes the value of $f_2^{(n)}(1)$ as

$$f_2^{(n)}(1) = \psi_1 \tag{7.13}$$

In the same way boundary conditions are establish at right contact, i.e. for $i = N$

$$a_2(N) = c_2(N) = 0, \quad \text{and} \quad b_2(N) = 1 \tag{7.14}$$

and

$$f_2^{(n)}(N) = \psi_N \tag{7.15}$$

Implementation of method 2 is the same as for method 1 and the results are the same. Here we only show results generated by method 2.

7.3 Solution and Results

The above problem is solved using Matlab functions. Using the files listed in Table 7.2 we analyzed p-n junction for step doping.

We have tested our program for several doping values:

$$N_A = 10^{22}m^{-3}, \ N_D = 10^{22}m^{-3}$$
$$N_A = 10^{22}m^{-3}, \ N_D = 10^{24}m^{-3}$$
$$N_A = 10^{24}m^{-3}, \ N_D = 10^{24}m^{-3}$$

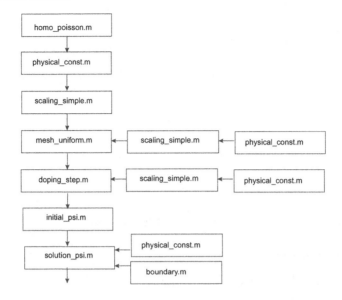

FIGURE 7.1
Flow diagram for homo diode with step doping (program *homo_poisson.m*).

We found reasonable agreement with other results [2, 3]. Results reported below are for the following doping: $N_A = 1 \times 10^{16} cm^{-3}$, $N_D = 1 \times 10^{17} cm^{-3}$. Potential is shown in Fig. 7.3, electric field is shown in Fig. 7.4 and densities of electrons and holes are shown in Fig. 7.5. Final results, e.g. total charge vs position strongly depend on the initial doping.

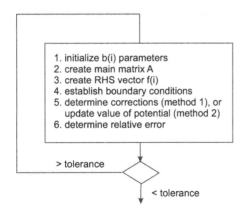

FIGURE 7.2
Flow diagram for function *solution_psi.m*.

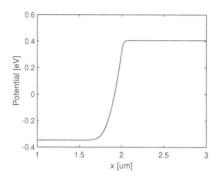

FIGURE 7.3
Potential of homo-junction vs position at equilibrium.

Appendix 7A: MATLAB Listings

Listing 7.A.1.1
Function determines values of potential (ψ) at contacts assuming ohmic contacts. It is used by two methods discussed in this chapter. Different boundary conditions are easily introduced by modifying it.

```
function [psi_L, psi_R] = boundary_psi(N_mesh,dop)
% File name: boundary_psi.m
% Provides values of potential at contacts without bias
% Ohmic contacts are assumed
% Contacts are numbered as i=0 (left) and i=N+1 (right)
% for method 1
% Contacts are numbered as i=1 (left) and i=N (right) for
% method 2
```

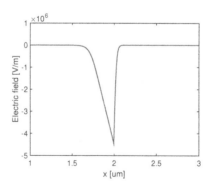

FIGURE 7.4
Electric field profile of homo-junction vs position at equilibrium.

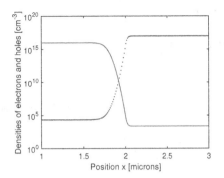

FIGURE 7.5
Electron and hole densities of homo-junction vs position at equilibrium.

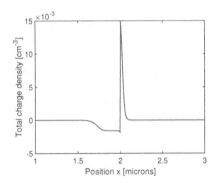

FIGURE 7.6
Total charge density of homo-junction vs position at eqilibrium.

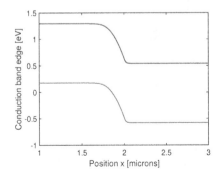

FIGURE 7.7
Conduction and valence band edges for homo-junction in equilibrium.

```
% In method 1, for practical reasons left contact has the
% same doping
% as mesh point i=1 and right contact has doping the same
% as mesh
% point i=N

% left contact: i = 0 (method 1) and i = 1 (method 2)
a_1 = 0.5*dop(1);
if (a_1>0)
    temp_1  = a_1*(1.0 + sqrt(1.0+1.0/(a_1*a_1)));
    elseif(a_1<0)
        temp_1 = a_1*(1.0 - sqrt(1.0+1.0/(a_1*a_1)));
end
psi_L = log(temp_1);

% right contact: i = N_mesh+1 (method 1) and i = N (method 2)
a_N = 0.5*dop(N_mesh);
if (a_N>0)
    temp_N  = a_N*(1.0 + sqrt(1.0+1.0/(a_N*a_N)));
    elseif(a_N<0)
        temp_N = a_N*(1.0 - sqrt(1.0+1.0/(a_N*a_N)));
end
psi_R = log(temp_N);
```

Listing 7.A.1.2
Function determines doping and scales it.

```
function dop = doping_step(N_mesh)
% Set up the doping and scale it with C_0

scaling_simple;
% Doping values in engineering units
N_A = 1d16;                % [cm^-3]
N_D = 1d17;                % [cm^-3]
% Convert to SI system
N_A = N_A*1d6;             % [m^3]
N_D = N_D*1d6;             % [m^3]

for i = 1:N_mesh
    if(i <= N_mesh/2)
        dop(i) = - N_A/C_0;
    elseif(i > N_mesh/2)
        dop(i) = N_D/C_0;
    end
end

end
```

Listing 7.A.1.3
Driver for both methods.

```
% File name: homo_poisson.m
% 1D model of homogeneous p-n junction based on Poisson equation
% Performs calculations for p-n junction in equilibrium

close all
clear all
tic

physical_const;     % Needed for n_int
scaling_simple;

[N_mesh,dx,x_plot]=mesh_uniform(); % outputs are scaled; x_plot
        % in microns
        % x_plot contains only coordinates if internal points,
        % not contacts
dop = doping_step(N_mesh);          % output is scaled
psi_trial = initial_psi(N_mesh,dop); % initial values only for device
                                    % output values are scaled
% determines potential in equilibrium
%psi = solution_psi_1(N_mesh,dx,psi_trial,dop); % using method 1
psi = solution_psi_2(N_mesh,dx,psi_trial,dop);  % using method 2

for i = 2:N_mesh-1
    rho(i) = -C_0*(exp(psi(i)) - exp(-psi(i)) - dop(i));
    % total charge density
    el_field(i) = -(psi(i+1) - psi(i))*V_T/(dx*x_0);  % electric field
end
%
for i = 1:N_mesh
    n(i) = exp(psi(i));
    p(i) = exp(-psi(i));
end

el_field(1) = el_field(2);
el_field(N_mesh) = el_field(N_mesh-1);
rho(1) = rho(2); rho(N_mesh) = rho(N_mesh-1);

toc

% Calculations of band edges
for i =2:N_mesh-1
    E_c(i) = E_0 - chi_e - V_T*psi(i);
    E_v(i) = E_0 - chi_e - E_g - V_T*psi(i);
end
E_c(1) = E_c(2); E_c(N_mesh) = E_c(N_mesh -1);
E_v(1) = E_v(2); E_v(N_mesh) = E_v(N_mesh -1);
```

```
% redefine quantities for plotting
n_plot = n*C_0*1d-6;  % rescale and convert to cm^-3
p_plot = p*C_0*1d-6;

psi_plot = V_T*psi;
rho_plot = q*rho*1d-6;
%
plot(x_plot, psi_plot,'LineWidth',1.5)
xlabel('x [um]','Fontsize',16);
ylabel('Potential [eV]','Fontsize',16);
set(gca,'FontSize',14);          % size of tick marks on both axis
%title('Potential vs position at equilibrium');
pause
%
plot(x_plot, el_field,'LineWidth',1.5)
xlabel('x [um]','Fontsize',16);
ylabel('Electric field [V/m]','Fontsize',16);
set(gca,'FontSize',14);          % size of tick marks on both axis
%title('Field profile vs position at equilibrium');
pause
%
semilogy(x_plot,n_plot,'.',x_plot,p_plot,'LineWidth',1.5);
xlabel('Position x [microns]','Fontsize',16);
ylabel('Densities of electrons and holes [cm^{-3}]','Fontsize',16);
set(gca,'FontSize',14);          % size of tick marks on both axis
pause
%
plot(x_plot,rho_plot,'LineWidth',1.5);
xlabel('Position x [microns]','Fontsize',16);
ylabel('Total charge density [cm^{-3}]','Fontsize',16);
set(gca,'FontSize',14);          % size of tick marks on both axis
pause
%
plot(x_plot,E_c,x_plot,E_v,'LineWidth',1.5);
xlabel('Position x [microns]','Fontsize',16);
ylabel('Conduction band edge [eV]','Fontsize',16);
set(gca,'FontSize',14);          % size of tick marks on both axis
pause

close all
```

Listing 7.A.1.4
Function initializes potential.

```
function psi = initial_psi(N_mesh,dop)
% Initialize potential and densities based on the requirement
% of charge
% neutrality throughout the structure, but not at contacts

for i = 1: N_mesh
```

```
    zz = 0.5*dop(i);
    if(zz > 0)
        xx = zz*(1 + sqrt(1+1/(zz*zz)));
    elseif(zz <  0)
        xx = zz*(1 - sqrt(1+1/(zz*zz)));
    end
    psi(i) = log(xx);
end

end
```

Listing 7.A.1.5
Function creates one-dimensional mesh

```
function [N_mesh,dx,x_plot] = mesh_uniform()
% Creates 1D uniform mesh

scaling_simple;

% Definition of structure
a = 1.0;             % left coordinate of junction; unit [micron]
b = 3.0;             % right coordinate of junction; unit [microns]
N_mesh = 200;        % number of mesh points

% x_plot -  mesh for plotting; contains only coordinates of
% internal points
dx = (b-a)/(N_mesh+1);  % mesh size [microns]
x_plot(1) = a;       % units [microns]
for m = 1:N_mesh
    x_plot(m) = a + m*dx;
end
%
dx = dx*1d-6;        % Convert mesh size to meters
dx = dx/x_0;         % Scale mesh size (value 1.5523 after scaling)
end
```

Listing 7.A.1.6
Summarizes physical constants.

```
% File name: physical_const.m
% Definitions of basic physical constants

eps_zero   = 8.85d-12;  % Permittivity of free space [F/m]
q      = 1.60d-19;      % elementary charge [C]
k_B    = 1.38d-23;      % Boltzmann constant [J/K]
```

```
T     = 300;              % Temperature [K]
%-------------------------
% Material's parameters
n_int   = 1.5d16;         % Intrinsic carrier concentration [m^-3]
eps_r = 11.7;             % Relative dielectric constant for Si
%
E_0 = 5;                  % Reference energy [eV]
chi_e = 4.05;             % Electron affinity for silicon [eV]
E_g = 1.12;               % Bandgap of silicon [eV]
```

Listing 7.A.1.7
Summarizes simple scaling.

```
% File name: scaling_simple.m
% Purpose: keeps general and electrical scaling factors

physical_const;          % needed for n_int

V_T = k_B*T/q;
L_Dint = sqrt(eps_zero*V_T/(q*n_int));  % intrinsic Debye length

x_0 = L_Dint;            % unit [meters]
C_0 = n_int;             % unit [m^-3]
```

Listing 7.A.2.1
Solves for potential using method 1.

```
function psi = solution_psi_1(N_mesh,dx,psi,dop)
% File name: solution_psi_1.m
% Performs calculations for p-n junction in equilibrium
% Method 1
% Determines psi (potential) in the iteration process using
% Matlab
% functions

physical_const;          % needed for eps_r
dx2 = dx*dx;
error_iter = 0.0001;
relative_error = 1.0;    % Initial value of the relative error

while (relative_error >= error_iter)
    for i = 1:N_mesh
        b(i)=-(2.0+(dx2/eps_r)*(exp(-psi(i))+exp(psi(i)))));
    end

    for i=1:N_mesh
```

```
        A(i,i) = b(i);                % diagonal elements
    end
    for i=1:N_mesh-1
        A(i,i+1) = 1.0;               % above diagonal
    end
    for i=1:N_mesh-1
        A(i+1,i) = 1.0;               % below diagonal
    end
%-------------------------------------
% Creation of rhs vector
%     f = zeros(1,N_mesh);
% Initialize array to hold rhs vector
    for i=2:N_mesh-1
        aa = dop(i)+exp(-psi(i))-exp(psi(i));
        f(i)=-psi(i-1)+2.0*psi(i)-psi(i+1)-(dx2*aa)/eps_r;
    end
%-----------------------------------------
% Establish boundary conditions
[psi_L, psi_R] = boundary_psi(N_mesh,dop);
% values of psi at contacts
a_L = dop(1)+exp(-psi(1))-exp(psi(1));
a_R = dop(N_mesh)+exp(-psi(N_mesh))-exp(psi(N_mesh));
%
f(1)=-psi_L+2.0*psi(1)-psi(2)-(dx2*a_L)/eps_r;
f(N_mesh)=-psi(N_mesh-1)+2.0*psi(N_mesh)-psi_R-(dx2*a_R)/eps_r;

%----------------------------------------------------------------
    delta_psi(1:N_mesh) = f/A;   % Solve for correction potential
                                 % using Matlab routines
%----------------------------------------------------------------
 psi = psi + delta_psi;                % New value of potential
 relative_error = max(abs(delta_psi(1,1:N_mesh)./
 psi(1,1:N_mesh)));
%      count = count +1;

end                                 % end of while loop


end                    % end of function
```

Listing 7.A.3.1
Solves for potential using method 2.

```
function psi_1 = solution_psi_2(N_mesh,dx,psi,dop)
% File name: solution_psi_2.m
% Performs calculations for p-n junction in equilibrium
% Method 2
```

```
% Determines psi (potential) in the iteration process using
% Matlab
% functions

physical_const;      % needed for eps_r
dx2 = dx*dx;

k_iter= 0;
error_iter = 0.0001;
relative_error = 1.0;

while(relative_error >= error_iter)
    for i = 1: N_mesh
        a(i) = 1;
        c(i) = 1;
        b(i) = -(2+(dx2/eps_r)*(exp(psi(i))+exp(-psi(i))));
        f(i)=(dx2/eps_r)*(exp(psi(i))-exp(-psi(i))-...
            dop(i)-psi(i)*(exp(psi(i))+exp(-psi(i))));
        end
% Establishing boundary conditions
a(1) = 0; c(1) = 0; b(1) = 1;
a(N_mesh) = 0; c(N_mesh) = 0; b(N_mesh) = 1;
%
% Establishing boundary conditions for potential at contacts
[psi_L, psi_R] = boundary_psi(N_mesh,dop);

f(1) = psi_L;
f(N_mesh) = psi_R;
%
% Creation of main matrix
for i=1:N_mesh
    A(i,i) = b(i);               % diagonal elements
end
%
for i=1:N_mesh-1
    A(i,i+1) = c(i);             % above diagonal
    A(i+1,i) = a(i+1);            % below diagonal
end

k_iter = k_iter + 1;

ff=f';

psi_1 = A\ff;
psi_1 = psi_1';
```

```
delta_psi = psi_1 - psi;
relative_error = max(abs(delta_psi)./psi_1);

psi = psi_1;

end      % end of while loop
k_iter

end      % end of function
```

Bibliography

[1] Vasileska D., Goodnick S.M., and Klimeck G. *Computational Electronics.* CRC Press, Boca Raton, 2010.

[2] P. Nayak. 1D drift diffusion simulator for modeling pn-junction diode. Semiconductor Process/Device Simulation (EEE533). Arizona State University, 2008.

[3] https://nanohub.org/.

8

Hetero-Junction Using Poisson Equation

We use previously developed methods to numerically solve Poisson equation for hetero p-n diode.

8.1 Heterostructure Diode with Step Doping

In practical devices material parameters depend on the position. In the case of Poisson equation relative dielectric constant ε_r is that quantity. In the following we dropped primes indicating scaling. Thus, by assumption, all quantities are scaled.

In this chapter we describe simulations of two hetero-structures described by Yang et al [1].

8.2 Summary of Implemented Equations

For convenience, previously derived equations (see Chapter 6) are are summarized here. We use Method two, i.e. with contacts inside.The discretized Poisson equation is expressed in a condensed form as

$$a_2(i)\psi_{i-1}^{(n+1)} - b_2(i)\psi_i^{(n+1)} + c_2(i)\psi_{i+1}^{(n+1)} = f_2(i) \tag{8.1}$$

where the coefficients are defined as

$$a_2(i) = \frac{\varepsilon_{i-1}}{x_i - x_{i-1}} \tag{8.2}$$

$$b_2(i) = \bar{b}_2(i) + \Delta x_i \left(e^{-\psi_i^{(n)}} + e^{\psi_i^{(n)}} \right) \tag{8.3}$$

$$c_2(i) = \frac{\varepsilon_{i+1}}{x_{i+1} - x_i} \tag{8.4}$$

DOI: 10.1201/9781003265849-8

$$f_2(i) = -\Delta x_i \left[e^{-\psi_i^{(n)}} - e^{\psi_i^{(n)}} + C_{dop}(x_i) \right] - \left[a(i)\psi_{i-1}^{(n)} - \bar{b}(i)\psi_i^{(n)} + c(i)\psi_{i+1}^{(n)} \right] \tag{8.5}$$

and

$$\bar{b}_2(i) = \frac{\varepsilon_i}{x_{i+1} - x_i} + \frac{\varepsilon_i}{x_i - x_{i-1}} \tag{8.6}$$

The above equations are supplemented by boundary conditions. Close to contacts the above equations become

$$a_2(1)\psi_L^{(n+1)} - b_2(1)\psi_1^{(n+1)} + c_2(1)\psi_2^{(n+1)} = f_2(1) \tag{8.7}$$

and

$$a_2(N)\psi_{N-1}^{(n+1)} - b_2(N)\psi_N^{(n+1)} + c_2(N)\psi_R^{(n+1)} = f_2(N) \tag{8.8}$$

where

$$a_2(1) = \frac{\varepsilon_1}{x_1 - a_L}, \quad a_2(N) = \frac{\varepsilon_{N-1}}{x_N - x_{N-1}} \tag{8.9}$$

$$c_2(1) = \frac{\varepsilon_2}{x_2 - x_1}, \quad c_2(N) = \frac{\varepsilon_N}{b_R - x_N} \tag{8.10}$$

$$\bar{b}_2(1) = \frac{\varepsilon_1}{x_2 - x_1} + \frac{\varepsilon_1}{x_1 - a_L}, \quad \bar{b}_2(N) = \frac{\varepsilon_N}{b_R - x_N} + \frac{\varepsilon_N}{x_N - x_{N-1}} \tag{8.11}$$

and

$$f_2(1) = -\Delta x_1 \left[e^{-\psi_1^{(n)}} - e^{\psi_1^{(n)}} + C_{dop}(x_1) \right] - \left[a(1)\psi_L^{(n)} - \bar{b}(1)\psi_1^{(n)} + c(1)\psi_2^{(n)} \right] \tag{8.12}$$

$$f_2(N) = -\Delta x_N \left[e^{-\psi_N^{(n)}} - e^{\psi_N^{(n)}} + C_{dop}(x_N) \right]$$
$$- \left[a(N)\psi_{N-1}^{(n)} - \bar{b}(N)\psi_N^{(n)} + c(N)\psi_R^{(n)} \right] \tag{8.13}$$

Here a_L is the coordinate of left contact, b_R is the coordinate of the right contact and $\psi_L^{(n+1)}$ and $\psi_R^{(n+1)}$ are the values of potential at left and right contacts.

8.2.1 Nonuniform system (heterostructure)

We analyze structure discussed by K. Yang et al. [1]. It is a GaAs/AlGaAs n-N isotype abrupt heterostructure as shown in Table 8.1. As a test we also simulated uniform structure based on GaAs.

8.2.2 Description of functions

Flow diagram for simulation of heterostructure in equilibrium is shown in Fig. 8.1. Condensed list of functions is summarized in Table 8.2. Below is a more detailed description. Now, we will provide a short description of functions used:

TABLE 8.1

Structure discusssed by Yang et al [1]

Material	Thickness (μm)	Doping (cm^{-3})
$n - GaAs$	$15\mu m$	10^{17}
$N - Al_{0.25}Ga_{0.75}As$	$15\mu m$	10^{14}
$p - GaAs$	$20\mu m$	10^{17}

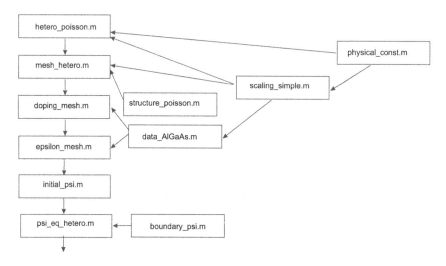

FIGURE 8.1

Flow diagram for heterostructure in equilibrium.

- boundary.m.

  This function provides values of potential and also densities of electrons and holes (for future use) at contacts. Ohmic contacts are assumed. Contacts are numbered as $i = 0$ (left) and $i = N + 1$ (right). All variables used in this function are scaled.

- data_AlGaAs.m

  This function provides data for AlGaAs used in the program. The values are summarized in Table 8.3. The used composition of Al was $x = 0.25$.

- doping_mesh.m

  Assigns values of doping for all mesh points.

- doping_values.m

- epsilon_mesh.m

  Assigns values of dielectric constant for all mesh points.

- hetero_poisson.m

TABLE 8.2
List of Matlab functions for Chapter 8. Heterojunction in equilibrium analyzed using Poisson equation.

Listing	Function	Description
	Data functions	
8.A.1.1	*data_AlGaAs.m*	Data for AlGaAs
8.A.1.2	*physical_const.m*	Inputs physical constants
8.A.1.3	*scaling_simple.m*	Provides scaling
8.A.1.4	*structure_poisson.m*	Defines analyzed structure
	Calculations	
8.A.2.1	*boundary_psi.m*	Boundary conditions for potential
8.A.2.2	*doping_mesh.m*	Parameters for step doping
8.A.2.3	*epsilon_mesh.m*	Generates mesh for heterostructure
8.A.2.4	*hetero_poisson.m*	Driver
8.A.2.5	*initial_psi.m*	Initializes potential
8.A.2.6	*mesh_hetero.m*	Creates simple 1D mesh
8.A.2.7	*psi_poisson_hetero.m*	Determines potential ψ
	Test data	
8.A.3.1	*data_AlGaAs_test.m*	data for AlGaAs homo
8.A.3.2	*structure_poisson_test*	structure for AlGaAs homo

TABLE 8.3
List of AlGaAs parameters.

Symbol	Value	Description
ε	$13.1 - 3x$	Dielectric constant
E_0	$5eV$	Reference energy
χ_e	$4.07 - 1.1x$	Electron affinity
E_g	$1.424 + 1.427x$	Bandgap
	0.67	Band offset

Driver. Solves one-dimensional model of p-n hetero-junction without bias voltage.

- initial_psi.m

  Assigns initial values of potential at mesh points for step doping.

- mesh_hetero.m

  Creates one-dimensional mesh for heterostructures.

- physical_const.m

  Defines basic physical constants.

- scaling_simple.m

  Keeps general scaling factors.

- psi_poisson_hetero.m

  Determines potential.

- structure_poisson.m

  Provides input to construct mesh.

8.2.3   Results for homo-structure

Results of simulations for this chapter are shown in Figs. 8.2, 8.3, and 8.4.

The above program was also tested assuming AlGaAs-homo diode. The results are shown in Figs.8.5, 8.6, and 8.7.

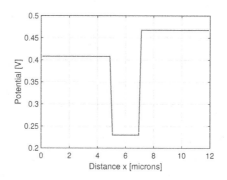

FIGURE 8.2

Potential vs position for 3-layer heterostructure.

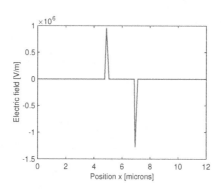

FIGURE 8.3

Electric field vs position for 3-layer heterostructure.

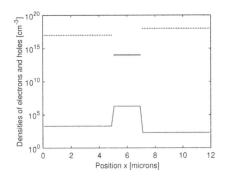

FIGURE 8.4
Densities vs position for 3-layer heterostructure.

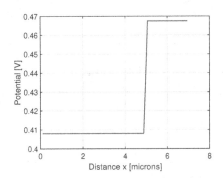

FIGURE 8.5
Potential vs position for 2-layer homostructure.

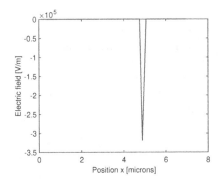

FIGURE 8.6
Electric field vs position for 2-layer homostructure.

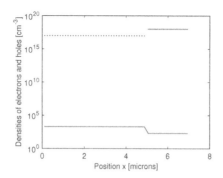

FIGURE 8.7
Densities vs position for 3-layer homostructure.

Appendix 8A: MATLAB Listings

8.2.4   Data functions

Listing 8.A.1.1

```
%-----------------------------------------------------------------
% File name: data_AlGaAS.m
% Structure based on: K. Yang et al, Solid State Elect.36,
% 321 (19930
%-----------------------------------------------------------------
scaling_simple;
%-----------------------------------------------------------------
% Aluminum composition in different layers
% Al_x Ga_1-x As
%-----------------------------------------------------------------
%comp_Al_layer(2)  = 0.0;   % This layer GaAs
%comp_Al_layer(1)  = 0.25;  %
x_comp = 0.25;
%
%-----------------------------------------------------------------
% Values of dielectric constant in each layer
% epsilon = (13.1 - 3x)*eps_zero
%-----------------------------------------------------------------
eps_layer(3)  = 13.1;
eps_layer(2)  = 13.1;
eps_layer(1)  = 13.1 - 3*x_comp;
```

```
%-------------------------------------------------------------
E_0 = 5;                              % reference energy [eV]
chi_e_GaAs = 4.07;
% Electron affinity for GaAs [eV]
E_g_GaAs = 1.42;                      % Bandgap of GaAs [eV]
chi_e_AlGaAs = 4.07-1.1*x_comp;
% Electron affinity for AlGaAs [eV]
E_g_AlGaAs = 1.424+1.427*x_comp;      % Bandgap of AlGaAs [eV]
band_offset = 0.67;
%----------------------------
chi_layer(3) = chi_e_GaAs;
chi_layer(2) = chi_e_AlGaAs;
chi_layer(1) = chi_e_GaAs;
E_g_layer(3) = E_g_GaAs ;
E_g_layer(2) = E_g_AlGaAs;
E_g_layer(1) = E_g_GaAs ;
%-------------------------------------------------------------
% Doping values in engineering units
N_A = 1d17;              % acceptor doping [cm^-3]
N_D = 1d18;              % donor doping [cm^-3]
N_neut = 1d14;           % undoped region [cm^-3]
% Convert to SI system
N_A = N_A*1d6;                % [m^3]
N_D = N_D*1d6;                % [m^3]
N_neut = N_neut*1d6;          % [m^-3]
%-------------------------------------------------------------
dop_layer(3) = N_D;       % in m^-3
dop_layer(2) = N_neut;        % in m^-3
dop_layer(1) = N_A;       % in m^-3
%------------- Scaling of data -----------------------
dop_layer = dop_layer/C_0;
```

Listing 8.A.1.2

```
% File name: physical_const.m
% Definitions of basic physical constants

eps_zero = 8.8541878d-12;
% permittivity of free space;  unit [C^2/Nm^2]
q = 1.6021892d-19;            % electron's charge; unit [Coulombs]
k_B = 1.380662d-23;           % Boltzman constant; unit [J/K]
T = 300;                      % absolute temperature; unit [K]
V_T = k_B*T/q;
%----------------------------
% Material's parameters
```

```
n_int = 1.4d16;              % intrinsic concentration; unit [m^-3]
```

Listing 8.A.1.3

```
% File name: scaling_simple.m
% Purpose: keeps general and electrical scaling factors

physical_const;          % needed for V_T, n_int

L_Dint = sqrt(eps*V_T/(q*n_int));

x_0 = L_Dint/5;              % unit [meters]
C_0 = n_int;                 % unit [m^-3]  -- critical parameter
V_T = k_B*T/q;
```

Listing 8.A.1.4

```
% File name: structure_poisson.m
% Input needed to construct mesh
% Structure based on: K. Yang et al, Solid State Elect.36, 321
% (1993)
%-----------------------------------------------------------------
% Values of layer thicknesses are establishd; values in microns
d_layer(3)  = 5.0;  % microns
d_layer(2)  = 2.0;  % microns
d_layer(1)  = 5.0;  % microns
%---- convert to meters ---------------------
d_layer = d_layer*1d-6;      % in meters
%-----------------------------------------------------------------
% Data needed to construct mesh
%-----------------------------------------------------------------
% N_mesh_layer(k) - number of mesh points in layer 'k'
%-----------------------------------------------------------------
N_mesh_layer(3)   = 40;
N_mesh_layer(2)   = 30;
N_mesh_layer(1)   = 40;
%
a_L = 0;    % arbitrary selected location of left contact
b_R = a_L + sum(d_layer);   % location of right contact
```

8.2.5 Calculations

Listing 8.A.2.1

```
function [psi_L, psi_R] = boundary_psi(N,dop)
% File name: boundary_psi.m
% Provides values of potential at contacts
% Ohmic contacts are assumed
% Contacts are numbered as i=0 (left) and i=N+1 (right)
% For practical reasons left contact has the same doping as mesh
% point i=0
% and right contact has doping same as mesh point i=N+1

% left contact: i = 0
a_1 = 0.5*dop(1);
if (a_1>0)
    temp_1 = a_1*(1.0 + sqrt(1.0+1.0/(a_1*a_1)));
    elseif(a_1<0)
        temp_1 = a_1*(1.0 - sqrt(1.0+1.0/(a_1*a_1)));
end
psi_L = log(temp_1);

% right contact: i = N+1
a_N = 0.5*dop(N_mesh);
if (a_N>0)
    temp_N = a_N*(1.0 + sqrt(1.0+1.0/(a_N*a_N)));
    elseif(a_N<0)
        temp_N = a_N*(1.0 - sqrt(1.0+1.0/(a_N*a_N)));
end
psi_R = log(temp_N);
```

Listing 8.A.2.2

```
function dop_mesh = doping_mesh(N,interface)
%--------------------------------------------------------------------
% File name: doping_mesh.m
% Purpose:   Assigns the value of doping for all mesh points.
%            That value stays the same within each layer and changes
%            from layer to layer.
% N                 - total number of mesh points
% interface(n)      - number of mesh points in layers, from the
%                     beginning to layer 'n'
% dop_mesh          - keeps doping for each mesh point

dop_mesh=zeros(1,N); % Initializes output array to zero
%-------------------------------
% layer - index layer
% Within a given layer, doping_layer is assigned the same value
% of doping.
```

```
% Loop scans all mesh points.
% For all mesh points selected for a given layer, the same
% value of doping is assigned.
%---------------------------------
data_AlGaAs;         % needed for dop_layer
%data_AlGaAs_test;
layer = 1;
for i = 1:N
   if i <= interface(layer)
      dop_mesh(i) = dop_layer(layer); % assigns value of epsilon
   else
      dop_mesh(i) = dop_layer(layer+1);
   end
   if i == interface(layer)+1
      layer = layer + 1;
   end
end
end
```

Listing 8.A.2.3

```
function eps_mesh = epsilon_mesh(N,interface)
%-------------------------------------------------------------------------
% File name: epsilon_mesh.m
% Purpose:   Assigns the value of dielectric constant for all
% mesh points.
%            That value stays the same within each layer and changes
%            from layer to layer.
% N               - total number of mesh points
% interface(n)    - number of mesh points in layers, from
% the beginning
%                     to layer 'n'
% eps_layer - dielectric constant
% eps_mesh        - keeps epsilon_dc for each mesh point

eps_mesh=zeros(1,N); % Initializes output array to zero

%---------------------------------
% layer - index layer
% Within a given layer, epsilon_layer is assigned the same value
% of dc dielectric constant.
% Loop scans all mesh points.
% For all mesh points selected for a given layer, the same
% value of eps_layer is assigned.
%---------------------------------


data_AlGaAs;         % needed for eps_layer
%data_AlGaAs_test;
layer = 1;
```

```
for i = 1:N
   if i <= interface(layer)
      eps_mesh(i) = eps_layer(layer); % assigns value of epsilon
   else
      eps_mesh(i) = eps_layer(layer+1);
   end
   if i == interface(layer)+1
      layer = layer + 1;
   end
end
```

Listing 8.A.2.4

```
% File name: hetero_poisson.m
% Solves one-dimensional model of p-n junction without bias voltage.
% Step doping is assumed
% All variables are scaled
%
clear all
close all
tic

physical_const;
scaling_simple;
%
V_b = 0;                        % Bias voltage

[x,x_layer,interface,N,a_L,b_R] = mesh_hetero();

dop = doping_mesh(N,interface);
eps = epsilon_mesh(N,interface);
psi_trial = zeros(1,N);    % Initialize array to hold trial potential
delta_psi = zeros(1,N);    % Initialize array to hold corrections
psi_trial = initial_psi(N, dop);
psi = psi_poisson_hetero(V_b,N,x,psi_trial,dop,eps,a_L,b_R);

toc

plot(x_0*x*1.0d6,V_T*psi,'LineWidth',1.5);  % Plot final result
% of iterations 1.0d6 is used to convert distance from meters to
% microns
xlabel('Distance x [microns]','Fontsize',16);
ylabel('Potential [V]','Fontsize',16);
set(gca,'FontSize',14);            % size of tick marks on both axis
grid on;

pause

% Determines electric field
dx = x(2)-x(1);        % temporary
```

```
for i=2:N-1
    electric_field(i)=-(psi(i+1)-psi(i))/(x(i+1)-x(i));
    n(i) = exp(psi(i));
    p(i) = exp(-psi(i));
end
n = n*C_0;
p = p*C_0;

electric_field(1)=electric_field(2);
electric_field(N)=electric_field(N-1);
%
electric_field = electric_field*V_T/x_0;      % rescale electric field

plot(x_0*x*1.0d6,electric_field,'LineWidth',1.5);
xlabel('Position x [microns]','Fontsize',14);
ylabel('Electric field [V/m]','Fontsize',14);
set(gca,'FontSize',14);              % size of tick marks on both axis
pause
%
n(1) = n(2); n(N) = n(N-1);
p(1) = p(2); p(N) = p(N-1);
semilogy(x_0*x*1.0d6,n*1d-6,'.',x_0*x*1.0d6,p*1d-6,'LineWidth',1.5);
xlabel('Position x [microns]','Fontsize',14);
ylabel('Densities of electrons and holes [cm^{-3}]','Fontsize',14);
set(gca,'FontSize',14);              % size of tick marks on both axis
pause

close all
```

Listing 8.A.2.5

```
function psi_init = initial_psi(N, dop)
% File name: initial_psi.m
% Assign initial values at internal mesh points for step doping.
% Initialize the potential based on the requirement of charge
% neutrality throughout the whole structure

for i = 1: N
    zz = 0.5*dop(i);
    if(zz > 0)
        xx = zz*(1 + sqrt(1+1/(zz*zz)));
    elseif(zz <  0)
        xx = zz*(1 - sqrt(1+1/(zz*zz)));
    end
    psi_init(i) = log(xx);
end

end
```

Listing 8.A.2.6

```
function [x,x_layer,interface,N,a_L,b_R] = mesh_hetero()
% Creates 1D mesh for heterostructures
%------------------------------
% Output
% x_layer      - right coordinates of layers
% x            - coordinates of mesh poinst
% interface(n) - total number of mesh points up to layer 'n'
% N            - total number of mesh points
% a_L          - location of left contact
% b_R          - location of right contact
%-----------------------------------------------
%structure_poisson_test;
structure_poisson;       % Data needed to construct mesh
                % d_layer         - thicknesses of each layer
                % N_mesh_layer(k) - number of mesh points in layer 'k
scaling_simple;

N_layers = length(d_layer);  % determine number of layers

for k = 1:N_layers
    delta_x(k)=d_layer(k)/(N_mesh_layer(k)+1); % dx in each layer
end

for m = 1: N_mesh_layer(1)
    x_1(m) = m*delta_x(1);          % mesh coordinates in first layer
end
%
% thickness up to layer 'k'
thickness(1) = d_layer(1);
for m = 2:N_layers
    thickness(m) = thickness(m-1) + d_layer(m);
end

for k = 2:N_layers               % loop over all layers
    clear x_t
    for m = 1:N_mesh_layer(k)    % loop within layer 'k'
        x_t(m) =  m*delta_x(k); % mesh points for layer 'k'
    end
        x_t = thickness(k-1) + x_t;
        u = [x_1, x_t];
        x_1 =u;
end
x = x_1;    % rename variables

%%%%%%%%%%%%%%%%%%%%%%%%%%%%%%%%%%%%%%%%%%%%%%%%%%%%%%%%%%%%%
N = length(x);   % total number of mesh points
%----------------------------------------------------------------
% Determine right coordinates of all layers
```

```
x_layer(1) = d_layer(1);
%
for k = 2:N_layers % loop over all layers
   x_layer(k) = x_layer(k-1) + d_layer(k);
end
%------------------------------------------------------------
% Create vector interface(k) which keeps number of mesh points up to
% a layer 'k'
interface(1) = N_mesh_layer(1);
for j = 1:N_layers-1
   interface(j+1) = interface(j)+N_mesh_layer(j+1);
end
%---- scaling for output
x = x/x_0;
x_layer = x_layer/x_0;
a_L = a_L/x_0;
b_R = b_R/x_0;
end
```

Listing 8.A.2.7

```
function psi = psi_poisson_hetero(V_b,N,x,psi,dop,ep,a_L,b_R)

% Determines psi (potential) in the iteration process using Matlab
% functions
% Discretization -> linearization
% Contacts outside

error_iter = 0.0001;            % iteration error
relative_error = 1.0;            % Initial value of the relative error

%-------------------------------
% Creation of matrix A

while (relative_error >= error_iter)
    for i = 2:N-1
        dx(i) = (x(i+1)-x(i-1))/2;
    end
    dx(1) = x(1)-a_L;
    dx(N) = b_R - x(N);

    for i = 2:N-1
        a(i) =  ep(i-1)/(x(i)-x(i-1));
        c(i) =  ep(i+1)/(x(i+1)-x(i));
        b_bar(i) = ep(i)/(x(i+1)-x(i))+ep(i)/(x(i)-x(i-1));
        b(i) =  b_bar(i) + dx(i)*(exp(V_b-psi(i))+exp(psi(i)));
    end
    a(1) = ep(1)/(x(1)-a_L);
    a(N) = ep(N)/(b_R-x(N));
```

```
    c(1) = ep(1)/(x(1)-a_L);
    c(N) = ep(N)/(b_R-x(N));

    b_bar(1) = ep(1)/(x(2)-x(1))+ep(1)/(x(1)-a_L);
    b(1) = b_bar(1) + dx(1)*(exp(V_b-psi(1))+exp(psi(1)));

    b_bar(N)=ep(N)/(b_R-x(N))+ep(N)/(x(N)-x(N-1));
    b(N) = b_bar(N) + dx(N)*(exp(V_b-psi(N))+exp(psi(N)));

    for i=1:N
        A(i,i) = - b(i);            % diagonal elements
    end
    for i=1:N-1
        A(i,i+1) = a(i);      % above diagonal
    end
    for i=1:N-1
        A(i+1,i) = c(i);  % below diagonal
    end
%-----------------------------------
% Creation of rhs vector
%     f = zeros(1,N_mesh);            % Initialize array to hold rhs vector
    for i=2:N-1
        aa = exp(-psi(i)) - exp(psi(i)) + dop(i);
        f(i) = -dx(i)*aa - a(i)*psi(i-1)+b_bar(i)*psi(i)-c(i)*psi(i+1);
    end
%---------------------------
% Establish boundary conditions for RHS
[psi_L, psi_R] = boundary_psi(N,dop);  % values of psi at contacts
aa_L = exp(-psi(1))-exp(psi(1)) + dop(1);
aa_R = exp(-psi(N))-exp(psi(N)) + dop(N);
%
f(1)= -(dx(1)*aa_L) - a(1)*psi_L + b_bar(1)*psi(1)-c(1)*psi(2);
f(N)=-(dx(N)*aa_R)-a(N)*psi(N-1)+b_bar(N)*psi(N)-c(N)*psi_R;
%-----------------------------------------
delta_psi(1:N) = f/A;
psi = psi + delta_psi;
relative_error = max(abs(delta_psi(1,1:N)./psi(1,1:N)));
end                 % end of while loop

end                 % end of function
```

8.2.6   Test data

Listing 8.A.3.1

```
% File name: data_AlGaAs_test.m
%-------------------------------------------------------------
%physical_const;    % needed for n_int
```

```
%doping_values;
scaling_simple;
%-------------------------------------------------------------
% Aluminum composition in different layers
% Al_x Ga_1-x As
%-------------------------------------------------------------
%comp_Al_layer(2)  = 0.0;    % This layer GaAs
%comp_Al_layer(1)  = 0.25;   %
x_comp = 0.25;
%
%-------------------------------------------------------------
% Values of dielectric constant in each layer
% epsilon = (13.1 - 3x)*eps_zero
%-------------------------------------------------------------
%eps_layer(3)  = 13.1;
eps_layer(2)   = 13.1;
eps_layer(1)   = 13.1; % - 3*x_comp;
%-------------------------------------------------------------
E_0 = 5;                          % reference energy [eV]
chi_e_GaAs = 4.07;                % Electron affinity for GaAs [eV]
E_g_GaAs = 1.42;                  % Bandgap of GaAs [eV]
chi_e_AlGaAs = 4.07-1.1*x_comp;   % Electron affinity for AlGaAs [eV]
E_g_AlGaAs = 1.424+1.427*x_comp;  % Bandgap of AlGaAs [eV]
band_offset = 0.67;
%----------------------------
%chi_layer(3) = chi_e_GaAs;
chi_layer(2) = chi_e_AlGaAs;
chi_layer(1) = chi_e_GaAs;
%E_g_layer(3) = E_g_GaAs ;
E_g_layer(2) = E_g_AlGaAs;
E_g_layer(1) = E_g_GaAs ;
%-------------------------------------------------------------
% Doping values in engineering units
N_A = 1d17;              % acceptor doping [cm^-3]
N_D = 1d18;              % donor doping [cm^-3]
N_neut = 1d14;           % undoped region [cm^-3]
% Convert to SI system
N_A = N_A*1d6;           % [m^3]
N_D = N_D*1d6;           % [m^3]
N_neut = N_neut*1d6;     % [m^-3]
%-------------------------------------------------------------
dop_layer(2) = N_D;         % in m^-3
%dop_layer(2) = N_neut;      % in m^-3
dop_layer(1) = N_A;         % in m^-3
%------------- Scaling of data -------------------
dop_layer = dop_layer/C_0;
```

Listing 8.A.3.2

```
% File name: structure_poisson_test.m
% Input needed to construct mesh
%----------------------------------------------------------------
% Values of layer thicknesses are establishd; values in microns
%d_layer(3)  = 5.0;  % microns
d_layer(2)  = 2.0;  % microns
d_layer(1)  = 5.0;  % microns
%---- convert to meters ---------------------
d_layer = d_layer*1d-6;      % in meters
%----------------------------------------------------------------
% Data needed to construct mesh
%----------------------------------------------------------------
% N_mesh_layer(k) - number of mesh points in layer 'k'
%----------------------------------------------------------------
%N_mesh_layer(3)  = 40;
N_mesh_layer(2)  = 30;
N_mesh_layer(1)  = 40;
%
a_L = 0;    % arbitrary selected location of left contact
b_R = a_L + sum(d_layer);   % location of right contact
```

Bibliography

[1] K. Yang, J.R. East, and G.I. Haddad. Numerical modeling of abrupt heterojunctions using a thermionic-field emission boundary condition. *Solid
State Electron.*, 36:321–330, 1993.

9

Homo-Diode Based on Drift-Diffusion

In this chapter, we will discuss about the advance model of p-n diode developed in the previous chapter by adding carrier transport equations to the Poisson equation already existing. Carrier transport of electrons and holes is described by drift-diffusion model. We also include fundamental generation-recombination processes. Together they will provide realistic model of p-n junction and will be starting point for semiconductor laser model developed later. We concentrate on one-dimensional (1-D) case. Here, we also restrict ourselves to homogeneous material.

Some of the variables change by many orders of magnitude. It is therefore a common practice to introduce appropriate scaling. After discussing it, other choice of variables will be reviewed. Lets start first with the summary of basic electrical equations in 1D along x-axis.

9.1 Electrical Equations

Continuity equation and expression for current for electrons are [1] (assuming that $G_{gen} = 0$)

$$\frac{\partial n}{\partial t} = \frac{1}{q}\frac{\partial J_n}{\partial x} - R \qquad (9.1)$$

$$J_n = -qn\mu_n\frac{\partial \psi}{\partial x} + qD_n\frac{\partial}{\partial x}n \qquad (9.2)$$

Corresponding equations for holes are

$$\frac{\partial p}{\partial t} = -\frac{1}{q}\frac{\partial J_p}{\partial x} - R \qquad (9.3)$$

$$J_p = -qp\mu_p\frac{\partial \psi}{\partial x} - qD_p\frac{\partial}{\partial x}p \qquad (9.4)$$

In this chapter we assume only SRH recombination.

DOI: 10.1201/9781003265849-9

TABLE 9.1
Main parameters in SRH recombination.

$\tau_{n0}[s]$	$N_{n,ref}[cm^{-3}]$	$\tau_{p0}[s]$	$N_{p,ref}[cm^{-3}]$
5×10^{-5}	5×10^{16}	5×10^{-5}	5×10^{16}

9.1.1 SRH recombination

The Shockley-Read-Hall (SRH) recombination is assumed to be [2–4]

$$R = \frac{pn - n_{int}^2}{\tau_p (n + n_t) + \tau_n (p + p_t)} \tag{9.5}$$

where

$$n_t = n_{int} \exp(\frac{E_t - E_i}{V_T}), \quad p_t = n_{int} \exp(-\frac{E_t - E_i}{V_T}) \tag{9.6}$$

Here, E_t is the recombination level (trap energy) and E_i is the intrinsic Fermi level.

If the recombination-center atoms occupy a single energy level located exactly in the middle of the bandgap the above becomes [5]

$$R = \frac{pn - n_{int}^2}{\tau_p (n + n_{int}) + \tau_n (p + n_{int})} \tag{9.7}$$

where n_{int} is the intrinsic carrier density and τ_n and τ_p are the electron and hole lifetimes. Their values depend on the concentrations of dopands and are

$$\tau_n = \frac{\tau_n^0}{1 + \frac{N_A + N_D}{N_n^{ref}}}, \quad \tau_p = \frac{\tau_p^0}{1 + \frac{N_A + N_D}{N_p^{ref}}} \tag{9.8}$$

Parameters are summarized in a Table 9.1 [4].

9.1.2 Mobility models

We summarize basic models of mobility implemented here.

9.1.2.1 Field independent mobility

For the case when mobility does not depend on electric field we have used the following empirical expressions for phosphorous and boron doped silicon, following Vasileska [4]

$$\mu_n(N) = 68.5 + \frac{(1414 - 68.5)}{1 + (N/9.2 \times 10^{16})^{0.711}} \frac{cm^2}{V \ s} \tag{9.9}$$

$$\mu_p(N) = 44.9 + \frac{(470.5 - 44.9)}{1 + (N/2.23 \times 10^{17})^{0.719}} \frac{cm^2}{V \ s} \tag{9.10}$$

where $N = N_D + N_A$.

9.1.2.2 Field dependent mobility

Mobility models for Si are [6]

$$\mu(E) = \mu_0 \left[\frac{1}{1 + \left(\frac{\mu_0 E}{v_{sat}}\right)^\beta} \right]^{1/\beta} \tag{9.11}$$

where μ_0 is the low-field mobility, v_{sat} is a saturation velocity, E is electric field and β is a parameter equal to 1 (for electrons) and 2 (for holes).

We used the following expressions for temperature dependent low-field mobilities for Si [6]:

electrons

$$\mu_{0n} = \mu_{1n} + \frac{\mu_{2n}}{1 + \left(\frac{T}{300}\right)^{-3.8} \cdot \left(\frac{N}{1.072 \cdot 10^{17}}\right)^{0.73}} \quad \left[\frac{cm^2}{V \cdot s}\right] \tag{9.12}$$

where

$$\mu_{1n} = 55.24, \quad \mu_{2n} = 7.12 \cdot 10^8 \cdot T^{-2.3} - 55.24 \quad \left[\frac{cm^2}{V \cdot s}\right] \tag{9.13}$$

holes

$$\mu_{0p} = \mu_{1p} + \frac{\mu_{2p}}{1 + \left(\frac{T}{300}\right)^{-3.7} \cdot \left(\frac{N}{1.606 \cdot 10^{17}}\right)^{0.70}} \quad \left[\frac{cm^2}{V \cdot s}\right] \tag{9.14}$$

where

$$\mu_{1p} = 49.74, \quad \mu_{2p} = 1.35 \cdot 10^8 \cdot T^{-2.2} - 49.70 \quad \left[\frac{cm^2}{V \cdot s}\right] \tag{9.15}$$

Here N is the local (total) concentration of dopands in cm^{-3} and T is the temperature in degrees Kelvin. In this work we conducted our analysis at room temperature, i.e. $T = 300K$.

Saturation velocity for Si is modeled [6, 7]

$$v_n^{sat} = v_p^{sat} = \frac{2.4 \cdot 10^7}{1 + 0.8 \exp\left(\frac{T}{600}\right)} \quad \left[\frac{cm}{s}\right] \tag{9.16}$$

9.1.3 Boundary conditions

Boundary conditions are determined by the values at contacts. Contacts are selected at mesh points $i = 1$ (left contact) and $i = N$ (right contact), N is total number of mesh points. Such a choice was earlier named as Method 2. We start our discussion with determination of densities.

9.1.3.1 Densities

We assume that the device has ideal ohmic contacts at both electrodes where the excess carriers will be removed instantly, hence the charge neutrality will be maintained. In the thermal equilibrium, the boundary conditions for carrier concentrations are [2, 3] (no space charge), at the first (left) and the last (right) mesh points

$$p_1 - n_1 + C_{dop}(x_1) = 0 \tag{9.17}$$

$$p_N - n_N + C_{dop}(x_N) = 0 \tag{9.18}$$

and (assuming thermal equilibrium at both electrodes)

$$p_1 \cdot n_1 = n_{int}^2 \tag{9.19}$$

$$p_N \cdot n_N = n_{int}^2 \tag{9.20}$$

Solving the system of algebraic equations (9.17), (9.19) for the left contact gives

$$p_1 = -\frac{1}{2} C_{dop}(x_1) \left\{ 1 \pm \sqrt{1 + \left[\frac{2n_{int}}{C_{dop}(x_1)} \right]^2} \right\} \tag{9.21}$$

$$n_1 = \frac{n_{int}^2}{p_1} \tag{9.22}$$

$$n_N = \frac{1}{2} C_{dop}(x_N) \left\{ 1 \pm \sqrt{1 + \left[\frac{2n_{int}}{C_{dop}(x_N)} \right]^2} \right\} \tag{9.23}$$

$$p_N = \frac{n_{int}^2}{n_N} \tag{9.24}$$

9.1.3.2 Potential

Values of potential at contacts are important as they are affected by the external bias voltage V_{bias}. Through potential, V_{bias} affects densities of carriers. The values for potential are determined from relations [2]

$$n(x) = n_{int} e^{\frac{q}{k_B T} \psi(x)} \tag{9.25}$$

and

$$p(x) = n_{int} e^{\frac{q}{k_B T}(V_{bias} - \psi(x))} \tag{9.26}$$

From above, values of potential at contacts are

$$\frac{\psi_1}{V_T} = V_{bias} - \ln \frac{p_1}{n_{int}} \tag{9.27}$$

$$\frac{\psi_N}{V_T} = \ln \frac{n_N}{n_{int}} \tag{9.28}$$

where $V_T = \frac{k_B T}{q}$.

9.1.4 Trial values

The system of equations we attempt to solve is a nonlinear one. Good trial values (initial guesses) are required. The quality of initial guess plays significant role in the numerical procedure and strongly influence amount of time required to find solution. For this several methods have been invented [8].

Trial values for densities are determined assuming the space-charge neutrality condition. The trial values of carrier densities are those in thermal-equilibrium [2]. The values are determined at zero bias voltage. In this book we use the following method to determine trial values.

9.1.4.1 Trial values of densities

For two continuity equations, assuming space charge neutrality condition, for the first zero bias point we use the thermal equilibrium carrier densities [2]. The corresponding equations are

for p-region

$$p_{trial}(x) = -C_{dop}(x) \tag{9.29}$$

$$n_{trial}(x) = -\frac{n_{int}^2}{C_{dop}(x)} \tag{9.30}$$

for n-region

$$n_{trial}(x) = C_{dop}(x) \tag{9.31}$$

$$p_{trial}(x) = \frac{n_{int}^2}{C_{dop}(x)} \tag{9.32}$$

with $C_{dop}(x) = N_D^+(x) - N_A^-(x)$ being the doping profile.

9.1.4.2 Trial values of potential

Trial values for potential are determined based on Eqs. (9.27) and (9.28).

for p-region

$$\psi_{trial} = -\ln \frac{C_{dop}(x)}{n_{int}} \tag{9.33}$$

for n-region

$$\psi_{trial} = \ln \frac{C_{dop}(x)}{n_{int}} \tag{9.34}$$

TABLE 9.2

Possible choises of unknown variables.

1	2	3	4	5	6
ψ	ψ	ψ	ψ	ψ	ψ
p	$p' = \frac{p}{n_{in}}$	$\sigma_p = \log p'$	$\varphi_p = \sigma_p + \psi$	$\phi_p = \exp(\varphi_p)$	$p' = \exp(-\psi)\phi_p$
n	$n' = \frac{n}{n_{in}}$	$\sigma_n = \log n'$	$\varphi_n = \psi - \sigma_n$	$\phi_n = \exp(-\varphi_n)$	$n' = \exp(\psi)\phi_n$

9.1.5 Choice of electrical variables

Natural variables to use in simulations from the physics point of view are p, n, and ψ. However, there are many reasons (mostly numerical) to use other variables. One can observe that n and p exponentially depend on the potential ψ. That fact guided researchers to choose a new set of independent variables. The most popular choices are summarized in Table 9.2 (based on [8]).

The expressions for currents of holes and electrons are also provided by Polak et al [8] as well as the range of change for each set of variables. There will always be a trade-off between the exponential character of the unknown functions and the non-linearity of the equations. The variables φ_n and φ_p have a physical meaning and are known as the quasi-Fermi energies (levels). In this chapter our variables of choice are: p, n and ψ.

9.1.6 Summary of linearized Poisson equation

Linearization and discretization of Poisson equation was done previously, see Chapter 6. Here we also assumed uniform grid and homogeneous structure. Therefore ε_r remains constant over the whole structure. Previously established equations are summarized here for convenience (please observe change in notation).

Here we follow with the linearization developed in Method 2, see Chapter 6.

Potential ψ at mesh point i is expressed as

$$\psi_i^{(n+1)} = \psi_i^{(n)} + \delta\psi_i \tag{9.35}$$

where $\psi_i^{(n)}$ is the value from the previous step and $\delta\psi_i$ is a correction which obeys the following equation

$$a(i)\psi_{i-1}^{(n+1)} - b(i)\psi_i^{(n+1)} + c(i)\psi_{i+1}^{(n+1)} = f_i, \quad \text{for } i = 1, 2, ...N \tag{9.36}$$

where

$$a(i) = 1, \quad c(i) = 1 \tag{9.37}$$

$$b(i) = 2 + \frac{\Delta x^2}{\varepsilon_r}\left(e^{V_{bias} - \psi_i^{(n)}} + e^{\psi_i^{(n)}}\right) \tag{9.38}$$

The forcing function for the Poisson equation is

$$f_i = -\frac{\Delta x^2}{\varepsilon_r} \left[e^{V_{bias} - \psi_i^{(n)}} + e^{\psi_i^{(n)}} + C_{dop}(x_i) \right] - \psi_{i-1}^{(n)} + 2\psi_i^{(n)} - \psi_{i+1}^{(n)} \quad (9.39)$$

Values close to contacts are

$$f_1 = -\frac{\Delta x^2}{\varepsilon_r} \left[e^{V_{bias} - \psi_1^{(n)}} + e^{\psi_1^{(n)}} + C_{dop}(x_1) \right] - \psi_L^{(n)} + 2\psi_1^{(n)} - \psi_2^{(n)} \quad (9.40)$$

and

$$f_N = -\frac{\Delta x^2}{\varepsilon_r} \left[e^{V_{bias} - \psi_N^{(n)}} + e^{\psi_N^{(n)}} + C_{dop}(x_N) \right] - \psi_{N-1}^{(n)} + 2\psi_N^{(n)} - \psi_R^{(n)} \quad (9.41)$$

In the above $\psi_L^{(n)}$ and $\psi_R^{(n)}$ are values at left and right contacts, respectively.

9.2   Integration of Current Continuity Equation

The current continuity equations are based on drift-diffusion equations. When discretized using central difference schemes often oscillations are introduced, especially when the drift term is large. The problem can be solved by introducing sufficiently fine grid or by introduction additional dissipation. Those solutions can create additional problems, like extra storage necessary for the fine grids. Some other solutions are possible, the most popular one the Scharfetter-Gummel (S-G) scheme [9]. In the following we will illustrate the S-G scheme by deriving discretized form of current in one dimension. The axis is called $x - axis$. Main assumptions are:

1. current density is constant along the outer edges of dicretization volume. In 1D this means that current is the same at points x_i and x_{i+1}. We will label that current by its mit-point value between points x_i and x_{i+1} as $J_n \left(i + \frac{1}{2} \right)$ for electrons and $J_p \left(i + \frac{1}{2} \right)$ for holes, as shown in Fig. 9.1.

2. within the discretization volume i.e between points $x_{i-\frac{1}{2}}$ and $x_{i+\frac{1}{2}}$, electric field, mobility and generation-recombination (G-R) are also constant. Thus electron's mobility at point $x_{i+\frac{1}{2}}$ will be labeled as $\mu_n(i+\frac{1}{2})$ and electric field at point $x = x_{i+\frac{1}{2}}$ will be discretized as

$$E(x) = -\frac{d\psi}{dx} = -\frac{\psi_{i+1} - \psi_i}{x_{i+1} - x_i} \quad (9.42)$$

Referring to Fig. 9.1 and restricting to electron type of carriers, the one-dimensional expression for electric current of electrons; see (9.2) between nodes x_i and x_{i+1} takes the form

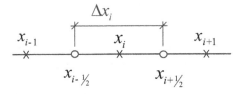

FIGURE 9.1
Details of one-dimensional mesh.

$$J_n\left(i + \frac{1}{2}\right) = -q\mu_n\left(i + \frac{1}{2}\right)n(x)\frac{d\psi(x)}{dx} + qD_n\left(i + \frac{1}{2}\right)\frac{dn(x)}{dx} \qquad (9.43)$$

This equation will now be integrated to determine electric current. First it is rewritten as an explicit differential equation for $n(x)$

$$\frac{dn(x)}{dx} - \frac{\mu_n\left(i + \frac{1}{2}\right)}{D_n\left(i + \frac{1}{2}\right)}\frac{\psi_{i+1} - \psi_i}{x_{i+1} - x_i}n(x) = \frac{1}{q}\frac{J_n\left(i + \frac{1}{2}\right)}{D_n\left(i + \frac{1}{2}\right)} \qquad (9.44)$$

Introduce notation

$$a = \frac{\mu_n\left(i + \frac{1}{2}\right)}{D_n\left(i + \frac{1}{2}\right)}\frac{\psi_{i+1} - \psi_i}{x_{i+1} - x_i} = \frac{\mu_n}{D_n}\left(i + \frac{1}{2}\right)\frac{\psi_{i+1} - \psi_i}{x_{i+1} - x_i} \qquad (9.45)$$

$$b = \frac{1}{q}\frac{J_n\left(i + \frac{1}{2}\right)}{D_n\left(i + \frac{1}{2}\right)} = \frac{1}{q}\frac{J_n}{D_n}\left(i + \frac{1}{2}\right) \qquad (9.46)$$

The ratio of a and b is evaluated to be

$$\frac{b}{a} = \frac{1}{q}\frac{J_n}{\mu_n}\left(i + \frac{1}{2}\right)\frac{x_{i+1} - x_i}{\psi_{i+1} - \psi_i} \qquad (9.47)$$

With the above definitions equation (9.44) takes the form

$$\frac{dn(x)}{dx} - a\,n(x) = b \qquad (9.48)$$

General solution of it is

$$n(x) = Ce^{ax} - \frac{b}{a} \qquad (9.49)$$

where C is a constant to be determined. Apply the above solution to mesh points x_i and x_{i+1} and obtain

$$n(x_i) = n_i = Ce^{ax_i} - \frac{b}{a}$$

$$n(x_{i+1}) = n_{i+1} = Ce^{ax_{i+1}} - \frac{b}{a}$$

Evaluating constant C from each of the above equations and comparing results, one finds

$$e^{-a\,x_i}\left(n_i + \frac{b}{a}\right) = e^{-ax_{i+1}}\left(n_{i+1} + \frac{b}{a}\right)$$

The ratio $\frac{b}{a}$ from above equation is

$$\frac{b}{a} = n_{i+1}\frac{e^{-ax_{i+1}}}{e^{-ax_i} - e^{-ax_{i+1}}} - n_i\frac{e^{-ax_i}}{e^{-ax_i} - e^{-ax_{i+1}}} \qquad (9.50)$$

Substituting result (9.50) into equation (9.47) one finds

$$\frac{J_n}{\mu_n}(i + \frac{1}{2}) = q\frac{\psi_{i+1} - \psi_i}{x_{i+1} - x_i}\left\{n_{i+1}\frac{1}{e^{a(x_{i+1}-x_i)} - 1} - n_i\frac{1}{1 - e^{-a(x_{i+1}-x_i)}}\right\} \qquad (9.51)$$

Term $a\,(x_{i+1} - x_i)$ is evaluated further to be

$$a\,(x_{i+1} - x_i) = \frac{\mu_n}{D_n}(i + \frac{1}{2})\,(\psi_{i+1} - \psi_i) = \frac{1}{V_T}\,(\psi_{i+1} - \psi_i) \qquad (9.52)$$

In the last step we have used Einstein relation (at point $i + \frac{1}{2}$)

$$D_n = \mu_n V_T \qquad (9.53)$$

where

$$V_T = \frac{k_B T}{q} \qquad (9.54)$$

Combining results (9.51) and (9.52) one finally finds the expression for current density for electrons at mid-point $x_{i+\frac{1}{2}}$

$$\frac{J_n}{\mu_n}(i + \frac{1}{2}) = q\frac{V_T}{x_{i+1} - x_i}\{n_{i+1}B(\frac{\psi_{i+1} - \psi_i}{V_T}) - n_i B(\frac{\psi_i - \psi_{i+1}}{V_T})\} \qquad (9.55)$$

In a similar way we can obtain expression for current density for holes at mid-point $x_{i+\frac{1}{2}}$

$$\frac{J_p}{\mu_p}(i + \frac{1}{2}) = q\frac{V_T}{x_{i+1} - x_i}\{-p_{i+1}B(\frac{\psi_i - \psi_{i+1}}{V_T}) + p_i B(\frac{\psi_{i+1} - \psi_i}{V_T})\} \qquad (9.56)$$

In the above expressions we used Bernoulli function is defined as

$$B(x) \equiv \frac{x}{e^x - 1} \qquad (9.57)$$

Various approximations to Bernoulli function are summarized next.

9.3 Approximations to Bernoulli Function

In numerical simulations it is necessary to have very accurate values of Bernoulli function and therefore its implementation requires care due to the possibility of division by zero. Another reason is that in semiconductor devices carrier concentrations can vary by more than 20 orders of magnitude. Typically [10] one approximates Bernoulli function in different ranges using different functions. One of the possible implementation is [11]

$$
B(x) = \begin{cases}
-x & x \leq -36.0 \\[2mm]
\frac{x}{\exp(x)-1} & -36.0 < x \leq -8.0 \times 10^{-7} \\[2mm]
1.0 - \frac{x}{2} & -8.0 \times 10^{-7} < x \leq 2.0 \times 10^{-7} \\[2mm]
\frac{x \exp(-x)}{1-\exp(-x)} & 2.0 \times 10^{-7} < x \leq 36.0 \\[2mm]
x \exp(x) & 36 < x \leq 746.0 \\[2mm]
0.0 & 746.0 < x
\end{cases}
\tag{9.58}
$$

Another approximation used by Vasileska et al [4] is

$$
B(x) = \begin{cases}
\frac{x \exp(-x)}{1-\exp(-x)} & 0.01 < x \\[2mm]
\frac{x}{\exp(x)-1} & x < 0 \\[2mm]
1 & x = 0
\end{cases}
\tag{9.59}
$$

The implementation used here follows Vasileska [4]. The above approximations ensures that no singularities will arise during the evaluation of $B(x)$. The Bernoulli function with the negative argument is evaluated as

$$
B(-x) = x + B(x)
\tag{9.60}
$$

9.4 Steady State: Discretization

From now on we work in a steady state, i.e.

$$
\frac{\partial n}{\partial t} = \frac{\partial p}{\partial t} = 0
\tag{9.61}
$$

9.4.1   Discretization of electrons and holes

Continuity equations in scaled form are

$$\frac{\partial J_n}{\partial x} - qR = 0 \tag{9.62}$$

$$\frac{\partial J_p}{\partial x} + qR = 0 \tag{9.63}$$

Discretization at point x_{i+1}(see Fig. 9.1) is done as

$$\frac{J_n(i+\frac{1}{2}) - J_n(i-\frac{1}{2})}{\Delta x_i} - qR_i = 0 \tag{9.64}$$

$$\frac{J_p(i+\frac{1}{2}) - J_p(i-\frac{1}{2})}{\Delta x_i} + qR_i = 0 \tag{9.65}$$

where $J_p(i+\frac{1}{2})$ is the value of hole current at intermediate point $x_{i+\frac{1}{2}}$ as given by Eq.(9.56).

Expressions for currents were derived before. Combining Eqs.(9.55) and (9.64) gives

$$\frac{\mu_n(i+\frac{1}{2})}{x_{i+1}-x_i}\left\{n_{i+1}B(\frac{\psi_{i+1}-\psi_i}{V_T}) - n_iB(\frac{\psi_i-\psi_{i+1}}{V_T})\right\} -$$
$$-\frac{\mu_n(i-\frac{1}{2})}{x_i-x_{i-1}}\left\{n_iB(\frac{\psi_i-\psi_{i-1}}{V_T}) - n_{i-1}B(\frac{\psi_{i-1}-\psi_i}{V_T})\right\}$$
$$= \frac{\Delta x_i}{V_T}R_i \tag{9.66}$$

In a similar way, combining Eqs.(9.56) and (9.65) gives for uniform mesh

$$\frac{\mu_p(i+\frac{1}{2})}{x_{i+1}-x_i}\left\{-p_{i+1}B(\frac{\psi_i-\psi_{i+1}}{V_T}) + p_iB(\frac{\psi_{i+1}-\psi_i}{V_T})\right\} -$$
$$-\frac{\mu_p(i-\frac{1}{2})}{x_i-x_{i-1}}\left\{-p_iB(\frac{\psi_{i-1}-\psi_i}{V_T}) + p_{i-1}B(\frac{\psi_i-\psi_{i-1}}{V_T})\right\}$$
$$= -\frac{\Delta x_i}{V_T}R_i \tag{9.67}$$

To arrive at the equations ready for numerical implementation the variables must be scaled.

9.5   Scaling

In the above equations, the dependent variables typically vary by many orders of magnitude on the points within solution domain. They also vary between themselves. Therefore, the dependent variables should be appropriately scaled according to their characteristic values. With the proper scaling, then the (dimensionless) dependent variables should be typically of the order of unity [1, 10]. The introduction of scaling will be discussed soon. Before, we summarize possible different choices of variables and then derive discretized expressions for electrical currents which also provides us with a possible way of how to introduce scaling.

As mentioned above, due to variation of dependent variables by many orders of magnitude, scaling plays an important role. For the scaling of the above equations we employ definitions as specified in Table 9.3 [1, 8, 10]. Definitions of scaling factors and their typical values are summarized in Table 9.4. For mobilities μ, recombination factors R, lifetimes τ_n, τ_p and time t we adopted here the same definitions of scaling factors as Kramers and Hitchon [1] (their Table 5.1) and also as Selberherr [10] (his Table 5.5).

TABLE 9.3
Definitions of general and electrical scaling variables.
Electric field E is scales 'naturally' as it is constructed
from already scaled variables ψ and x.

Name	Quantity	Definition of scaling
distance	x	$x' = \frac{x}{x_0}$
potential	ψ	$\psi' = \frac{\psi}{V_T}$
concentrations	n, p	$n' = \frac{n}{C_0}, p' = \frac{p}{C_0}$
doping	C_{dop}, n_{int}	$C'_{dop} = \frac{C_{dop}}{C_0}, n'_{int} = \frac{n_{int}}{C_0}$
time	t	$t' = \frac{t}{t_0}$
current	$J_{n,p}$	$J'_{n,p} = \frac{J_{n,p}}{J_0}$
diffusion	$D_{n,p}$	$D'_{n,p} = \frac{D_{n,p}}{D_0}$
mobility	$\mu_{n,p}$	$\mu'_{n,p} = \frac{\mu_{n,p}}{\mu_0}$
electric field	E	$E' = \frac{E}{E_0}$

TABLE 9.4

General and electrical scaling factors.

Note: For the practical implementation here we choose $C\_0 = n\_int$

Description	Symbol	Definition	Value	Reference
thermal voltage	V_T	$\frac{k_B T}{q}$	$0.0259 V$	[10]
length scale	x_0	N/A	λ_D	[8]
concentration	C_0	N/A	$1.5 \times 10^{16} m^{-3}$	[8]
diffusion scale	D_0	N/A	$10^{-4} m^2/s$	
Debye length	λ_D	$\frac{1}{x_0}\sqrt{\frac{V_T \varepsilon_0}{C_0 q}}$	1.86×10^{-10}	
time	t_0	$\frac{x_0^2}{D_0}$	$9.7 \times 10^{-7} s$	[1]
current	J_0	$\frac{x_0}{q D_0 C_0}$	$83 A cm^{-2}$	[1]
mobility	μ_0	$\frac{D_0}{V_T}$	$1000 cm^2/(Vs)$	[1]
electric field	E	$\frac{V_T}{x_0}$		

9.5.1   Scaling at boundaries

$$p'_1 = \frac{C'_{dop}(x_1)}{2}\left\{1 + \sqrt{1 + \left(\frac{2}{C'_{dop}(x_1)}\right)^2}\right\} \tag{9.68}$$

$$n'_1 = \frac{1}{p'_1} \tag{9.69}$$

$$p'_N = \frac{C'_{dop}(x_N)}{2}\left\{1 + \sqrt{1 + \left(\frac{2}{C'_{dop}(x_N)}\right)^2}\right\} \tag{9.70}$$

$$n'_N = \frac{1}{p'_N} \tag{9.71}$$

For $\frac{1}{C'_{dop}(x_1)} \ll 1$ and $\frac{1}{C'_{dop}(x_N)} \ll 1$ (which is always true except at very high temperatures), one has

$$p'_1 \approx C'_{dop}(x_1), \qquad p'_N \approx C'_{dop}(x_N) \tag{9.72}$$

The above expressions can also be expressed as

$$\psi'_1 = V'_{bias} - \ln p'_1 \tag{9.73}$$

$$\psi'_N = \ln n'_N \tag{9.74}$$

9.5.2   Scaling of trial values of potential

$$\psi'^{trial}_i(x_i) = \begin{cases} \ln\left\{\frac{n'_{int}}{C'_{dop}(x_i)}\right\}, & \text{for p-region} \\ \ln\left\{\frac{C'_{dop}(x_i)}{n'_{int}}\right\}, & \text{for n-region} \end{cases} \tag{9.75}$$

9.5.3 Scaling of mobilities

Scaling of mobilities is done as, see Table 9.3

$$\mu'_n = \frac{\mu_n}{\mu_0} \tag{9.76}$$

and

$$\mu'_p = \frac{\mu_p}{\mu_0} \tag{9.77}$$

Value of μ_0 is in a Table 9.4.

9.5.4 Scaling of recombination

R_i specifies all recombination terms at mesh point x_i. After scaling it has the form

$$R_i = R(i) = \frac{p_i n_i - 1}{\tau_p (n_i + 1) + \tau_n (p_i + 1)} \tag{9.78}$$

9.5.5 Scaling of continuity equations

9.5.5.1 Scaling of electrons

In the last step we will scale all relevant quantities. Final results are provided separately for electrons and holes.

In the next step we perform full scaling of Eqs.(9.66) and (9.67) by using definitions from Table 9.3. Applying scaling to variables in Eq.(9.66) one finds

$$\frac{\mu'_n(i + \frac{1}{2})}{x'_{i+1} - x'_i} \left\{ n'_{i+1} B(\psi'_{i+1} - \psi'_i) - n'_i B(\psi'_i - \psi'_{i+1}) \right\} -$$

$$-\frac{\mu'_n(i - \frac{1}{2})}{x'_i - x'_{i-1}} \left\{ n'_i B(\psi'_i - \psi'_{i-1}) - n'_{i-1} B(\psi'_{i-1} - \psi'_i) \right\}$$

$$= \frac{x_0}{\mu_0 C_0} \frac{\Delta x'_i}{V_T} x_0 \frac{C_0^2}{t_0 C_0} \frac{p'_i n'_i - 1}{\tau'_p (n'_i + 1) + \tau'_n (p'_i + 1)} \tag{9.79}$$

Using relations from Table 9.4 one finds for combination of scaling factors

$$\frac{x_0}{\mu_0 C_0} \frac{1}{V_T} x_0 \frac{C_0^2}{t_0 C_0} = \frac{x_0^2}{\mu_0 V_T t_0} = 1 \tag{9.80}$$

Therefore, the final expression used to determine densities of electrons on mesh points is (*we have dropped primes as all variables are now scaled*)

$$\frac{\mu_n(i + \frac{1}{2})}{x_{i+1} - x_i} \left\{ n_{i+1} B(\psi_{i+1} - \psi_i) - n_i B(\psi_i - \psi_{i+1}) \right\} -$$

$$-\frac{\mu_n(i - \frac{1}{2})}{x_i - x_{i-1}} \left\{ n_i B(\psi_i - \psi_{i-1}) - n_{i-1} B(\psi_{i-1} - \psi_i) \right\}$$

$$= \Delta x_i \frac{p_i n_i - 1}{\tau_p (n_i + 1) + \tau_n (p_i + 1)} \tag{9.81}$$

The above can be written in a matrix form as

$$a_n(i)n_{i-1} - b_n(i)n_i + c_n(i)n_{i+1} = f_n(i) \tag{9.82}$$

where

$$a_n(i) = \frac{\mu_n(i-\frac{1}{2})}{x_i - x_{i-1}} B\left(\psi_{i-1} - \psi_i\right) \tag{9.83}$$

$$b_n(i) = \frac{\mu_n(i+\frac{1}{2})}{x_{i+1} - x_i} B\left(\psi_i - \psi_{i+1}\right) + \frac{\mu_n(i-\frac{1}{2})}{x_i - x_{i-1}} B\left(\psi_i - \psi_{i-1}\right) \tag{9.84}$$

$$c_n(i) = \frac{\mu_n(i+\frac{1}{2})}{x_{i+1} - x_i} B\left(\psi_{i+1} - \psi_i\right) \tag{9.85}$$

The forcing function for electrons (recombination term)

$$f_n(i) = \Delta x_i \frac{p_i n_i - 1}{\tau_p\left(n_i + 1\right) + \tau_n\left(p_i + 1\right)} \tag{9.86}$$

9.5.5.2   Scaling of remaining quantities. Holes

In a similar way one finds for holes (all variables scaled)

$$a_p(i)p_{i-1} - b_p(i)p_i + c_p(i)p_{i+1} = f_p(i) \tag{9.87}$$

where

$$a_p(i) = \frac{\mu_p(i-\frac{1}{2})}{x_i - x_{i-1}} B\left(\psi_i - \psi_{i-1}\right) \tag{9.88}$$

$$b_p(i) = \frac{\mu_p(i+\frac{1}{2})}{x_{i+1} - x_i} B\left(\psi_{i+1} - \psi_i\right) + \frac{\mu_p(i-\frac{1}{2})}{x_i - x_{i-1}} B\left(\psi_{i-1} - \psi_i\right) \tag{9.89}$$

$$c_p(i) = \frac{\mu_p(i+\frac{1}{2})}{x_{i+1} - x_i} B\left(\psi_i - \psi_{i+1}\right) \tag{9.90}$$

and force function for holes

$$f_p(i) = \Delta x_i \frac{p_i n_i - 1}{\tau_p\left(n_i + 1\right) + \tau_n\left(p_i + 1\right)} \tag{9.91}$$

The above can be cleary expressed as tridiagonal matrices.

9.6   Electric Current

Electric current consists of two contributions, electrons and holes. Current due to electrons at the intermediate mesh point $i+\frac{1}{2}$ is (all quantities are not scaled)

$$J_n(i + \frac{1}{2}) = \mu_n(i + \frac{1}{2})q\frac{V_T}{x_{i+1} - x_i}\{n_{i+1}B(\frac{\psi_{i+1} - \psi_i}{V_T}) - n_iB(\frac{\psi_i - \psi_{i+1}}{V_T})\}$$

(9.92)

In the program we determine results in a scaled form, therefore current must be expressed in terms of scaled variables. Applying scaling one obtains (we preserved scaled notation)

$$J_n(i + \frac{1}{2}) = q\mu_0 qV_T\frac{C_0}{x_0[cm]}\mu'_n(i + \frac{1}{2})\frac{1}{x'_{i+1} - x'_i}\{n'_{i+1}B(\psi'_{i+1} - \psi'_i) - n'_iB(\psi'_i - \psi'_{i+1})\}$$

(9.93)

The prefactor P becomes (we have selected $C_0 = n_{int}$)

$$P = \frac{qV_Tn_{int}}{x_0[cm]}\mu_0 = \frac{qn_{int}}{x_0[cm]}D_0, \quad [\frac{C\ cm^{-3}}{cm}\frac{cm^2}{s} = \frac{A}{cm^2}]$$

(9.94)

In the same way one can express current of holes.

9.7 Results

In this section we provide results of the analysis of homo-diode and therefore some simplifications of the derived equations for electrons and holes are possible. For homogeneous case we have created homogeneous mesh and therefore

$$\Delta x = x_i - x_{i-1}$$

For illustration we simulated simple structure based on silicon. It is shown in Table 9.5. The structure is defined in the file *mesh_uniform.m*.

Matlab code for this simulation is provided in the next chapter. The results of calculations are shown in the following Figures.

9.7.1 Results at equilibrium

In Figs. 9.2, 9.3, 9.4 and 9.5 we plotted potential, electric field, densities of electrons and holes and also total charge density across the structure. The obtained results agree well with those published in [4] and [12].

TABLE 9.5
Simulated structure.

Left coordinate (μm)	Right coordinate (μm)	Number of mesh points
1.0	3.0	60

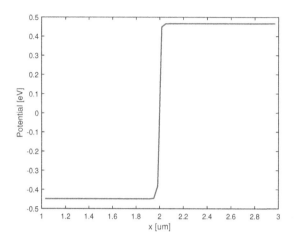

FIGURE 9.2
Potential in equilibrium for homo-junction.

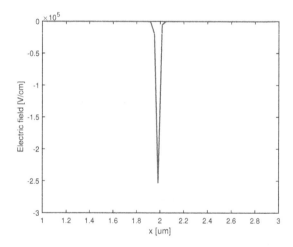

FIGURE 9.3
Electric field in equilibrium for homo-junction.

9.7.2   Results for non-equilibrium

Results for non-equilibrium case are shown in Figs.9.6, 9.7, 9.8 and 9.9. They
agree well with those published in [4] and [12].

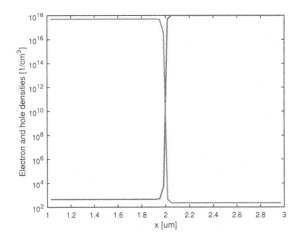

FIGURE 9.4
Densities of electrons and holes in equilibrium for homo-junction.

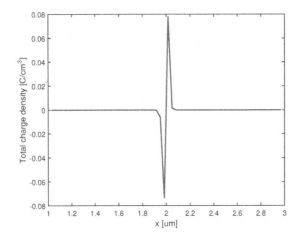

FIGURE 9.5
Total charge density in equilibrium for homo-junction.

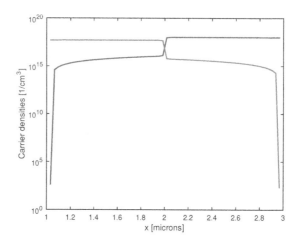

FIGURE 9.6
Densities of electrons and holes, non-equilibrium case.

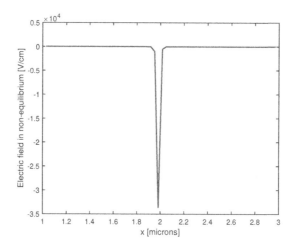

FIGURE 9.7
Electric field, non-equilibrium case.

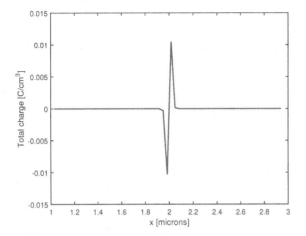

FIGURE 9.8
Total charge, non-equilibrium case

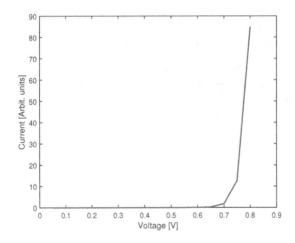

FIGURE 9.9
Current-voltage characteristic of the analyzed junction.

Bibliography

[1] K.M. Kramer and W.N.G. Hitchon. *Semiconductor Devices. A Simulation Approach.* Prentice Hall, Upper Saddle River, New Jersey, 1997.

[2] M. Kurata. *Numerical Analysis for Semiconductor Devices.* Lexington-Books, New York, 1982.

[3] M. Shur. *Physics of Semiconductor Devices*. Prentice Hall, Englewood Cliffs, NJ, 1990.

[4] Vasileska D., Goodnick S.M., and Klimeck G. *Computational Electronics*. CRC Press, Boca Raton, 2010.

[5] S.-L. Chuang. *Physics of Photonics Devices*. Wiley, New York, 2009.

[6] 1993. ATLAS II. User's Manual. Version 1.0.

[7] S.M. Sze. *Physics of Semiconductor Devices*. Wiley, New York, 1969.

[8] S.J. Polak, C. Den Heijer, and W.H.A. Schilders. Semiconductor device modelling from the numerical point of view. *International Journal for Numerical Methods in Engineering*, 24:763–838, 1987.

[9] D.L. Scharfetter and H.K. Gummel. Large-signal analysis of a silicon read diode oscillator. *IEEE Trans. Electron Devices*, 16:64–77, 1969.

[10] S. Selberherr. *Analysis and Simulation of Semiconductor Devices*. Springer, Wien, New York, 1984.

[11] A.W. Smith and K.F. Brennan. Hydrodynamic simulation of semiconductor devices. *Prog. Quant. Electr.*, 21:293–360, 1998.

[12] P. Nayak. 1D drift diffusion simulator for modeling pn-junction diode. Semiconductor Process/Device Simulation (EEE533). Arizona State University, 2008.

10

Matlab Code for p-n Homo-Diode

10.1 Summary of Implemented Equations: Homogeneous case

For convenience below we summarize main formulas which are used for Matlab implementation provided in this chapter. All equations are in a scaled form.

Equations for electrons and holes are provided below. They were established in the previous chapter.

1. Poisson equation at equilibrium

$$a_2(i)\psi_{i+1}^{(n+1)} + b_2(i)\psi_i^{(n+1)} + c_2(i)\psi_{i-1}^{(n+1)} = f^{(n)}(i) \qquad (10.1)$$

where

$$a_2(i) = 1, \quad b_2(i) = -\left[2 + \frac{\Delta x^2}{\varepsilon_r}\left(e^{-\psi_i^{(n)}} + e^{\psi_i^{(n)}}\right)\right], \quad c_2(i) = 1 \qquad (10.2)$$

and

$$f^{(n)}(i) = -\frac{\Delta x^2}{\varepsilon_r}\left[\left(e^{-\psi_i^{(n)}} - e^{\psi_i^{(n)}} + C_i\right) + \psi_i^{(n)}\left(e^{-\psi_i^{(n)}} + e^{\psi_i^{(n)}}\right)\right] \qquad (10.3)$$

2. Equation for electrons

$$a_n(i)n_{i-1} + b_n(i)n_i + c_n(i)n_{i+1} = f_{SRH}(i) \qquad (10.4)$$

where

$$a_n(i) = \mu_n(i - \frac{1}{2})B\left(\psi_{i-1} - \psi_i\right) \qquad (10.5)$$

$$b_n(i) = -\left[\mu_n(i - \frac{1}{2})B\left(\psi_i - \psi_{i-1}\right) + \mu_n(i + \frac{1}{2})B\left(\psi_i - \psi_{i+1}\right)\right] \qquad (10.6)$$

$$c_n(i) = \mu_n(i + \frac{1}{2})B\left(\psi_{i+1} - \psi_i\right) \qquad (10.7)$$

DOI: 10.1201/9781003265849-10

3. Equation for holes

$$a_p(i)p_{i-1} + b_p(i)p_i + c_p(i)p_{i+1} = f_{SRH}(i) \qquad (10.8)$$

where

$$a_p(i) = \mu_p(i - \frac{1}{2})B\left(\psi_i - \psi_{i-1}\right) \qquad (10.9)$$

$$b_p(i) = -\left[\mu_p(i - \frac{1}{2})B\left(\psi_{i-1} - \psi_i\right) + \mu_p(i + \frac{1}{2})B\left(\psi_{i+1} - \psi_i\right)\right]$$
$$(10.10)$$

$$c_p(i) = \mu_p(i + \frac{1}{2})B\left(\psi_i - \psi_{i+1}\right) \qquad (10.11)$$

4. SRH recombination

$$f_{SRH}(i) = \Delta x^2 \frac{p_i n_i - 1}{\tau_p\left(n_i + 1\right) + \tau_n\left(p_i + 1\right)} \qquad (10.12)$$

5. Poisson equation at non-equilibrium

$$a_2(i)\psi_{i+1}^{(n+1)} + b_2(i)\psi_i^{(n+1)} + c_2(i)\psi_{i-1}^{(n+1)} = f^{(n)}(i) \qquad (10.13)$$

where

$$a_2(i) = 1, \quad b_2(i) = -\left(2 + p_i^{(n)} + n_i^{(n)}\right), \quad c_2(i) = 1 \qquad (10.14)$$

and

$$f^{(n)}(i) = -\frac{\Delta x^2}{\varepsilon_r}\left[\left(p_i^{(n)} - n_i^{(n)} + C_i\right) + \psi_i^{(n)}\left(p_i^{(n)} + n_i^{(n)}\right)\right]$$
$$(10.15)$$

List of Matlab functions used to model homo-junction using drift-diffusion model in Table 10.1.

Appendix 10A: MATLAB Listings

10.1.1 Main functions

Listing 10.A.1.1 Implementation of Bernouli function.

```
function out = ber(x)
% File name: ber.m
% Implementation of Bernoulli function, based on
% Vasileska et al, Computational Electronics, CRC Press, 2010
```

TABLE 10.1
Homo-junction by drift-diffusion. List of Matlab files.

Listing	Function	Description
	Main functions	
10.A.1.1	*ber.m*	Implementation of Bernoulli function
10.A.1.2	*boundary_psi.m*	Boundary for potential (psi)
10.A.1.3	*currents.m*	Currents of electrons and holes
10.A.1.4	*diode_homo.m*	Homogeneous p-n junction
10.A.1.5	*doping_step.m*	Set up the doping
10.A.1.6	*electron_density.m*	Determines density of electrons
10.A.1.7	*hole_density.m*	Determines density of holes
10.A.1.8	*initial.m*	Initialize potential
10.A.1.9	*mesh_uniform.m*	Creates uniform mesh
10.A.1.10	*mobility_n_Si.m*	Mobility of electrons
10.A.1.11	*mobility_p_Si.m*	Mobility of holes
10.A.1.12	*potential_eq_1.m*	Solves Poisson's equation in equilibrium
10.A.1.13	*potential_noneq_2.m*	Solves Poisson's equation in non-equilibrium
10.A.1.14	*recombination_SRH.m*	Determines recombination of carriers
10.A.1.15	*solution_psi.m*	Solves for potential in equilibrium
	Definitions of parameters	
10.A.2.1	*material_parameters_Si.m*	Parameters for Si
10.A.2.2	*physical_const.m*	Physical constants
10.A.2.3	*scaling_dd.m*	Electrical scaling factors

```
flag_sum = 0;
        if(x>0.01)
           out = x*exp(-x)/(1-exp(-x));
        elseif(x<0&abs(x)> 0.01)
            out = x/(exp(x)-1);
        elseif(x == 0)
           out = 1;
        else
           temp_term = 1;
           sum = temp_term;
           i = 0;
           while(~flag_sum)
              i = i + 1;
              temp_term = temp_term*x/(i+1);
              if( sum + temp_term == sum)
                 flag_sum = 1;
              end
                 sum = sum + temp_term;
           end
           out = 1/sum;
        end
```

Listing 10.A.1.2
Boundary effects for electric potential (ψ).

```
function [psi_L, psi_R] = boundary_psi(N_mesh,dop)
% File name: boundary_psi.m
% Provides values of potential at contacts without bias
% Ohmic contacts are assumed
% Contacts are numbered as i=1 (left) and i=N (right) for
% method 2

a_1 = 0.5*dop(1);
if (a_1>0)
    temp_1  = a_1*(1.0 + sqrt(1.0+1.0/(a_1*a_1)));
    elseif(a_1<0)
        temp_1 = a_1*(1.0 - sqrt(1.0+1.0/(a_1*a_1)));
end
psi_L = log(temp_1);

% right contact: i = N_mesh+1 (method 1) and i = N (method 2)
a_N = 0.5*dop(N_mesh);
if (a_N>0)
    temp_N  = a_N*(1.0 + sqrt(1.0+1.0/(a_N*a_N)));
    elseif(a_N<0)
        temp_N = a_N*(1.0 - sqrt(1.0+1.0/(a_N*a_N)));
end
psi_R = log(temp_N);
```

Listing 10.A.1.3
File contains implementation of currents of electrons and holes.

```
function tot_curr_sum = currents(N,dx,dop,psi,n,p)

% File name: currents.m
% Calculates total current due to electrons and holes

tot_curr_sum = 0;
for i = 2:N-1

    [mu_min_n,mu_plus_n] = mobility_n_Si(i,dop,dx,psi);

    curr_n(i) = mu_plus_n*(n(i+1)*ber(psi(i+1) - psi(i))...
        - n(i)*ber(psi(i) - psi(i+1)));
        % Electron current density

    [rmu_min_p,rmu_plus_p] = mobility_p_Si(i,dop,dx,psi);
```

```
    curr_p(i) = rmu_plus_p*(p(i)*ber(psi(i+1)-psi(i))...
        - p(i+1)*ber(psi(i)-psi(i+1)));
        % Hole current density

    tot_curr(i) = curr_n(i) + curr_p(i);
    tot_curr_sum(i) = tot_curr_sum(i-1)+ tot_curr(i);
end

end
```

Listing 10.A.1.4
Driver for homo p-n junction.

```
% File name: diode_homo_nov.m
% 1D drift-diffusion model of homogeneous p-n junction
% November, 2022
% B.c. are introduced outside of computational domain
% for solving
% Poisson eq
% in equilibrium simuation
% For non-equilibrium calculations b.c. are introduced in the
% computational
% domain

clear all
close all

tic

'calculating'

scaling_dd;

[N,dx,x_plot] = mesh_uniform();
dop = doping_step(N);          % Assigns doping for each mesh point
                               % Scaling inside
[psi,n,p] = initial(N,dop); % initializes field, n, p
                               % No scaling inside
psi_eq = potential_eq_1(N,dx,psi,dop); % solution in equilibrium

for i = 2:N-1    % determines equilibrium quantities for plotting
    ro(i) = -n_int*(exp(psi_eq(i))-exp(-psi_eq(i))-dop(i));
    % total charge
    el_field(i) = -(psi_eq(i+1) - psi_eq(i))*V_T/(dx*x_0);
end
```

```matlab
for i = 1:N
    n(i) = exp(psi_eq(i));
    p(i) = exp(-psi_eq(i));
end

% Define values at boundaries (for plotting)
el_field(1) = el_field(2);
el_field(N) = el_field(N-1);
ro(1) = ro(2);
ro(N) = ro(N-1);

toc

% redefine equilibrium quantities for plotting
nf1 = n*n_int*1d-6;          % convert to cm^-3
pf1 = p*n_int*1d-6;          % convert to cm^-3
psi_eq_plot = V_T*psi_eq;    % potential in eV
el_field = el_field*1d-2;    % convert to V/cm
ro_f = q*ro*1d-6;            % convert to C/cm^3

% % %==== Plots of equilibrium values =========
% figure(1)
% plot(x_plot, psi_eq_plot,'LineWidth',1.5)
% xlabel('x [um]');
% ylabel('Potential [eV]');
% pause
%
% figure(2)
% plot(x_plot, el_field,'LineWidth',1.5)
% xlabel('x [um]');
% ylabel('Electric field [V/cm]');
% pause
%
% figure(3)
% semilogy(x_plot, nf,'LineWidth',1.5)
% hold on;
% semilogy(x_plot, pf,'LineWidth',1.5)
% xlabel('x [um]');
% ylabel('Electron and hole densities [1/cm^3]');
% pause
%
% figure(4)
% plot(x_plot, ro_f,'LineWidth',1.5)
% xlabel('x [um]');
```

```
% ylabel('Total charge density [C/cm^3]');
% pause
%
% close all
%-------------------------------------------------------------------
% Solving for the non-equillibirium case
delta_acc = 1E-7;              % Preset the tolerance

Va_max = 0.8;                  % max value of applied voltage [V]
dVa = 0.05;                    % value of step of applied voltage [V]
% Scale voltages
Va_max = Va_max/V_T;
dVa = dVa/V_T;

psi = psi_eq;

Va  = 0;
Va_max = Va_max;
if(abs(Va_max) == 0)
else
     k = 1;
    while(abs(Va) < abs(Va_max))
        Va = Va + dVa;
        psi(1) = psi(1) + dVa;
        flag = 0;
        k_iter = 0;
        while(~flag)    % convergence loop
            k_iter = k_iter + 1;
            n = electron_density_2(N,dx,dop,psi,n,p);
            p = hole_density_2(N,dx,dop,psi,n,p);
            [psi, delta_max] =
            potential_noneq_2(N,dx,dop,psi,n,p);
            if(delta_max < delta_acc)
                 flag = 1;
            end
        end             % End of the convergence loop (while loop)
%=================================================================

for i = 2:N-1    % determines electric field in non-equilibrium
    el_field_non_eq(i) = -(psi(i+1) - psi(i))*V_T/(dx*x_0);
end
el_field_non_eq(1) = el_field_non_eq(2);
el_field_non_eq(N) = el_field_non_eq(N-1);

% Calculates currents of electrons and holes
```

```
tot_curr_sum = currents(N,dx,dop,psi,n,p);
av_curr(k) = tot_curr_sum(N - 1)*1e-3/(N-2);

        for i = 2:N-1
            ro_non_eq(i) = - n_int* (n(i) - p(i) - dop(i));
            if(i>1)
            end
        end

        nf1 = n * n_int*1d-6;
        pf1 = p * n_int*1d-6;
        V(k) = Va*V_T;
        k = k + 1;
    end                    % end of voltage loop

aa = q*n_int*D_0/(dx*L_Dint*1d2);    % convert to A/cm^2

av_curr = aa*av_curr;

toc

el_field_non_eq = el_field_non_eq*1d-2;    % convert to V/cm
ro_non_eq = q*ro_non_eq*1d-6;              % convert to C/cm^3

figure(1);
semilogy(x_plot,nf1,'LineWidth',1.5);
% Plotting the final carrier densities
hold on;
semilogy(x_plot,pf1,'LineWidth',1.5);
xlabel('x [microns]');
ylabel('Carrier densities [1/cm^3]');
pause

figure(2);
plot(x_plot,el_field_non_eq,'LineWidth',1.5);
% Plotting the final potential
xlabel('x [microns]');
ylabel('Electric field in non-equilibrium [V/cm]');
pause

figure(3);
plot(x_plot(1:N-1),ro_non_eq,'LineWidth',1.5);
% Plotting the total charge
xlabel('x [microns]');
```

```
ylabel('Total charge [C/cm^3]');
pause

figure(4);
plot(V,av_curr,'LineWidth',1.5);   % Plotting the average current
xlabel('Voltage [V]');
ylabel('Current [Arbit. units]');
pause

close all

end
```

Listing 10.A.1.5
File defines step doping.

```
function dop = doping_step(N)
% Set up the doping and scale it with C_0

scaling_dd;
% Doping values in engineering units
N_A = 5d17;                 % [cm^-3]
N_D = 1d18;                 % [cm^-3]
% Convert to SI system
N_A = N_A*1d6;              % [m^3]
N_D = N_D*1d6;              % [m^3]

for i = 1:N
    if(i <= N/2)
        dop(i) = - N_A/C_0;
    elseif(i > N/2)
        dop(i) = N_D/C_0;
    end
end

end
```

Listing 10.A.1.6
Calculation of density of electrons.

```
function n = electron_density_2(N,dx,dop,psi,n,p)
% File name: electron_density_2.m
% Determines electron density using method 2

dx2 = dx*dx;
```

```
for i = 2: N-1
    [diff_min,diff_plus] = mobility_n_Si(i,dop,dx,psi);
    an(i) = diff_min*ber(psi(i-1) - psi(i));
    cn(i) = diff_plus*ber(psi(i+1) - psi(i));
    bn(i) = -(diff_min*ber(psi(i) - psi(i-1))
            + diff_plus*ber(psi(i)
            - psi(i+1)));
    fn(i) = dx2*recombination_SHR(i,dop,n,p);
end

an(1) = 0;
cn(1) = 0;
bn(1) = 1;
fn(1) = n(1);
an(N) = 0;
cn(N) = 0;
bn(N) = 1;
fn(N) = n(N);

N = N;
for i=1:N
    A(i,i) = bn(i);              % diagonal elements
end
for i=1:N-1
    A(i,i+1) = cn(i);               % above diagonal
end
for i=1:N-1
    A(i+1,i) = an(i+1);                % below diagonal
end

fn=fn';
n = A\fn;                  % 0.329684 seconds. good solution

end
```

Listing 10.A.1.7
Calculation of density of holes.

```
function p = hole_density_2(N,dx,dop,psi,n,p)
% File name: hole_density_2.m
% Determines hole density using method 2

dx2 = dx*dx;
```

```
for i = 2: N-1
    [diff_min,diff_plus] = mobility_p_Si(i,dop,dx,psi);
    ap(i) =  diff_min*ber(psi(i) - psi(i-1));
    cp(i) =  diff_plus*ber(psi(i) - psi(i+1));
    bp(i) = -(diff_min*ber(psi(i-1) - psi(i))
            + diff_plus*ber(psi(i+1)
            - psi(i)));
    fp(i) = dx2*recombination_SHR(i,dop,n,p);

end

ap(1) = 0;
cp(1) = 0;
bp(1) = 1;
fp(1) = p(1);
ap(N) = 0;
cp(N) = 0;
bp(N) = 1;
fp(N) = p(N);

for i=1:N
    A(i,i) = bp(i);             % diagonal elements
end
for i=1:N-1
    A(i,i+1) = cp(i);              % above diagonal
end
for i=1:N-1
    A(i+1,i) = ap(i+1);              % below diagonal
end

fp=fp';
p = A\fp;

end
```

Listing 10.A.1.8
Initializes potential.

```
function [psi,n,p] = initial(N,dop)
% Initialize potential based on the requirement of charge
% neutrality throughout the whole structure

for i = 1: N
    zz = 0.5*dop(i);
    if(zz > 0)
        xx = zz*(1 + sqrt(1+1/(zz*zz)));
```

```
    elseif(zz <   0)
        xx = zz*(1 - sqrt(1+1/(zz*zz))));
    end
    psi(i) = log(xx);
    n(i) = xx;
    p(i) = 1/xx;
end

end
```

Listing 10.A.1.9
Uniform mesh is created here.

```
function [N,dx,x_plot] = mesh_uniform()
% Creates 1D uniform mesh and scale it

scaling_dd;

% Definition of structure
a = 1.0;            % left coordinate of junction; unit [micron]
b = 3.0;            % right coordinate of junction; unit [microns]
N = 60;        % number of mesh points

% x_plot - mesh for plotting; contains only coordinates of
% internal
% points
dx = (b-a)/(N+1);   % mesh size [microns]
x_plot(1) = a;     % units [microns]
for m = 1:N
    x_plot(m) = a + m*dx;
end
%
dx = dx*1d-6;          % Convert mesh size to meters
dx = dx/x_0;           % Scale mesh size
end
```

Listing 10.A.1.10
Mobility of electrons in silicon.

```
function [mu_minus,mu_plus] = mobility_n_Si(i,dop,dx,psi)
% Determines mobilities of electrons for silicon at mesh
% point 'i'
% It is assumed that mobility depends on electric field
% Units of mobility [cm^2/(V s)
% Scaling is performed here
% Assumed room temperature, T=300K
```

```
scaling_dd;

mu_1n = 55.24;                        % [cm^2/(V s)]
mu_2n = 7.12*1d8*300^(-2.3) - 55.24;  % [cm^2/(V s)]
beta_n = 2;
vsat_n = 2.4e7/(1+0.8*exp(0.5));
% saturation velocity [cm/s]  for Si from Silvaco

N_dop = abs(dop(i));
aa = 1+ (N_dop/(1.072*1d17))^0.73;
% denominator for mu_n_zero
mu_n_zero = mu_1n + (mu_2n/aa);
% low-field electron mobility

E_plus = abs((psi(i)-psi(i+1)))/dx;
% electric field at mid-point above 'i'
E_minus  = abs((psi(i-1)-psi(i)))/dx;
% electric field at mid-point below 'i'

b_plus = mu_n_zero*E_plus/vsat_n;
dd = 1 + b_plus^beta_n;
mu_plus = mu_n_zero*((1/dd)^(1/beta_n));
mu_plus = mu_plus*1d-4;     % convert to [m^2/V s]

b_minus = mu_n_zero*E_minus/vsat_n;
dd = 1 + b_minus^beta_n;
mu_minus  = mu_n_zero*((1/dd)^(1/beta_n));
mu_minus = mu_minus*1d-4;   % convert to [m^2/V s]

mu_minus = mu_minus/mu_0;              % scaling
mu_plus = mu_plus/mu_0;

end
```

Listing 10.A.1.11
Mobility of holes in silicon.

```
function [mu_minus,mu_plus] = mobility_p_Si(i,dop,dx,psi)
% Determines mobilities of holes for silicon at mesh point 'i'
% It is assumed that mobility depends on electric field
% Units of mobility [cm^2/(V s)
% Scaling is performed here
% Assumed room temperature, T=300K

scaling_dd;
```

```
mu_1p = 49.74;                          % [cm^2/(V s)]
mu_2p = 1.35*1d8*300^(-2.2) - 49.70;   % [cm^2/(V s)]
beta_p = 1;
vsat_p = 2.4e7/(1+0.8*exp(0.5));
% saturation velocity [cm/s]  for Si from Silvaco

N_dop = abs(dop(i));
aa = 1+ (N_dop/(1.606*1d17))^0.70;
% denominator for mu_p_zero
mu_p_zero = mu_1p + (mu_2p/aa);
% low-field hole mobility

E_plus = abs((psi(i)-psi(i+1)))/dx;
% electric field at mid-point above 'i'
E_minus  = abs((psi(i-1)-psi(i)))/dx;
% electric field at mid-point below 'i'

b_plus = mu_p_zero*E_plus/vsat_p;
dd = 1 + b_plus^beta_p;
mu_plus = mu_p_zero*((1/dd)^(1/beta_p));
mu_plus = mu_plus*1d-4;    % convert to [m^2/V s]

b_minus = mu_p_zero*E_minus/vsat_p;
dd = 1 + b_minus^beta_p;
mu_minus  = mu_p_zero*((1/dd)^(1/beta_p));
mu_minus = mu_minus*1d-4;    % convert to [m^2/V s]

mu_minus = mu_minus/mu_0;                % scaling
mu_plus = mu_plus/mu_0;

end
```

Listing 10.A.1.12
Evaluates potential in equilibrium using Poisson equation.

```
function psi = potential_eq_1(N,dx,psi,dop)
% File name: solution_psi_1.m
% Performs calculations for p-n junction in equilibrium
% using method 1
% Determines psi (potential) in the iteration process
% using Matlab
% functions

material_parameters_Si;    % needed for eps_r
```

```
dx2 = dx*dx;

error_iter = 0.0001;
relative_error = 1.0;      % Initial value of the relative error

while (relative_error >= error_iter)
     for i = 1:N
         b(i)=-(2.0+(dx2/eps_r)*(exp(-psi(i))+exp(psi(i))));
     end

     for i=1:N
         A(i,i) = b(i);             % diagonal elements
     end
     for i=1:N-1
         A(i,i+1) = 1.0;               % above diagonal
     end
     for i=1:N-1
         A(i+1,i) = 1.0;               % below diagonal
     end
%------------------------------------
% Creation of rhs vector
%     f = zeros(1,N_mesh);
% Initialize array to hold rhs vector
     for i=2:N-1
         aa = exp(-psi(i))-exp(psi(i))+dop(i);
         bb = psi(i-1) - 2.0*psi(i) + psi(i+1);
         f(i)=-(dx2/eps_r)*aa - bb;
     end
%------------------------------------
% Establish boundary conditions
[psi_L, psi_R] = boundary_psi(N,dop);  % values of psi
% at contacts
a_L = exp(-psi(1))-exp(psi(1))+dop(1);
a_R = exp(-psi(N))-exp(psi(N)) + dop(N);
%
f(1)=-psi_L+2.0*psi(1)-psi(2)-(dx2*a_L)/eps_r;
f(N)=-psi(N-1)+2.0*psi(N)-psi_R-(dx2*a_R)/eps_r;

%---------------------------------------------------------------
     delta_psi(1:N) = f/A;  % Solve for correction potential
                            % using Matlab routines
%---------------------------------------------------------------
     psi = psi + delta_psi;                % New value of potential
     relative_error = max(abs(delta_psi(1,1:N)./psi(1,1:N)));
```

```
end                              % end of while loop

end              % end of function
```

Listing 10.A.1.13 Potential in non-equilibrium.

```
function [psi, delta_max] = potential_noneq_2(N,dx,dop,psi,n,p)
% File name: potential_noneq_2.m
% Solution of the Poisson's equation in non-equibrium
% Using method 2

material_parameters_Si;        % Needed for eos_r

delta_acc = 1E-7;              % Preset the tolerance

dx2 = dx*dx;
% Define coefficient of main matrix and RHS
for i = 2: N-1
    a(i) = 1;
    c(i) = 1;
    b(i) = -(2+(dx2/eps_r)*(p(i)+n(i)));
    f(i) = -(dx2/eps_r)*(p(i) - n(i) + dop(i) + psi(i)*(p(i)
           + n(i)));
end

% Initialize values at contacts
a(1) = 0;
c(1) = 0;
b(1) = 1;
f(1) = psi(1);
a(N) = 0;
c(N) = 0;
b(N) = 1;
f(N) = psi(N);

for i=1:N
    A(i,i) = b(i);               % diagonal elements
end

for i=1:N-1
    A(i,i+1) = c(i);                  % above diagonal
    A(i+1,i) = a(i+1);                % below diagonal
end

ff=f';
```

```
psi_1 = A\ff;
psi_1 = psi_1';

delta = psi_1 - psi;

psi = psi_1;

% Test update in the outer iteration loop

delta_max = 0;
for i = 1: N
    xx = abs(delta(i));
    if(xx > delta_max)
        delta_max=xx;
    end
end

end
```

Listing 10.A.1.14
SRH recombination.

```
function fn = recombination_SHR(i,dop,n,p)
% Determines recombination at mesh point 'i'

scaling_dd

tau_n0 = 1e-5;  % SHR recombination time for electrons
tau_p0 = 1e-5;  % SHR recombination time for holes
N_ref_n = 5e16;
N_ref_p = 5e16;

NN = 2*abs(dop(i));
tau_n = tau_n0/(1 + NN/N_ref_n);     % electron lifetime
tau_p = tau_p0/(1 + NN/N_ref_p);     % hole lifetime

rnum  = n(i)*p(i) - 1;
denom = tau_n*(p(i)+1)+tau_p*(n(i)+1);
denom = denom/t_0;       % scaling
fn    = (rnum/denom);
end
```

Listing 10.A.1.15
Solution for electrostatic potential, ψ.

```matlab
function psi_1 = solution_psi(N_mesh,dx,psi,dop)
% File name: solution_psi_2.m
% Performs calculations for p-n junction in equilibrium
% Method 2
% Determines psi (potential) in the iteration process using
% Matlab
% functions
physical_const;      % needed for eps_r
dx2 = dx*dx;

%k_iter= 0;
error_iter = 0.0001;
relative_error = 1.0;

while(relative_error >= error_iter)
    for i = 1: N_mesh
        a(i) = 1;
        c(i) = 1;
        b(i) = -(2+(dx2/eps_r)*(exp(psi(i))+exp(-psi(i))));
        f(i)=(dx2/eps_r)*(exp(psi(i))-exp(-psi(i))-...
            dop(i)-psi(i)*(exp(psi(i))+exp(-psi(i))));
    end
% Establishing boundary conditions
a(1) = 0; c(1) = 0; b(1) = 1;
a(N_mesh) = 0; c(N_mesh) = 0; b(N_mesh) = 1;
%
% Establishing boundary conditions for potential at contacts
[psi_L, psi_R] = boundary_psi(N_mesh,dop);

f(1) = psi_L;
f(N_mesh) = psi_R;
%
% Creation of main matrix
for i=1:N_mesh
    A(i,i) = b(i);            % diagonal elements
end
%
for i=1:N_mesh-1
    A(i,i+1) = c(i);                  % above diagonal
    A(i+1,i) = a(i+1);                % below diagonal
end

%k_iter = k_iter + 1;

ff=f';
```

```
psi_1 = A\ff;
psi_1 = psi_1';

delta_psi = psi_1 - psi;
relative_error = max(abs(delta_psi)./psi_1);

psi = psi_1;


end      % end while loop
%k_iter
end
```

10.1.2 Definitions of parameters

Listing 10.A.2.1
File contains parameters of silicon

```
% File name: material_parameters_Si.m
% Inputs neded material constants for silicon in SI system

eps_r = 11.7;           % Relative dielectric constant for Si
E_0 = 5;                % Reference energy [eV]
chi_e = 4.05;           % Electron affinity for silicon [eV]
E_g = 1.12;             % Bandgap of silicon [eV]
```

Listing 10.A.2.2
File contains physical constants.

```
% File name: physical_const.m
% Definitions of basic physical constants

eps_zero  = 8.85d-12;  % Permittivity of free space [F/m]
q     = 1.60d-19;      % elementary charge [C]
k_B   = 1.38d-23;      % Boltzmann constant [J/K]
T     = 300;           % Temperature [K]
n_int   = 1.5d16;      % Intrinsic carrier concentration [m^-3]
```

Listing 10.A.2.3
File contains scaling data.

```
% File name: scaling_dd.m
% Purpose: keeps general and electrical scaling factors
```

```
% for dd model
% Variable description:
% Basic scaling:
% V_T      - thermal voltage
% x_zero  - length scaling
% C_zero  - concentration scaling
% D_zero  - diffusion scaling
%
% Derived scaling:
% lambda  - Poisson equation scaling
% t_zero  - time scaling
% R_zero  - recombination scaling
% J_zero  - current scaling
% mu_zero - mobility scaling

physical_const;           % needed for V_T, n_int

V_T = k_B*T/q;            % Thermal voltage[eV]
L_Dint = sqrt(eps_zero*V_T/(q*n_int));

x_0 = L_Dint;             % unit [meters]
C_0 = n_int;              % unit [m^-3]
D_0 = 1d3;           % [cm^2/s]
t_0 = (x_0^2)/D_0;
mu_0 = D_0/V_T;
```

11

Hetero-Diode Based on Drift-Diffusion

Here we describe an implementation of drift-diffusion model for heterostructures. We start with the summary of relevant equations. I will use uniform mesh, i.e.

$$\Delta x = x_i - x_{i-1} = x_{i+1} - x_i \tag{11.1}$$

so all formulas are almost the same as drift-diffusion equations for homostructure with the observation that relative dielectric constant ε_r and mobilities μ_n, μ_p are position dependent. However, program which generate mesh used here needs modifications to account for different materials forming heterostructure.

We start with the summary of equations implemented in this chapter.

11.1   Poisson Equation in Equilibrium

In equilibrium Poisson equation involves only potential ψ. After linearization and discretization it can be expressed in a condensed form as

$$a_2(i)\psi_{i+1}^{(n+1)} + b_2(i)\psi_i^{(n+1)} + c_2(i)\psi_{i-1}^{(n+1)} = f^{(n)}(i) \tag{11.2}$$

where

$$a_2(i) = \varepsilon_r(i-1), \quad b_2(i) = -\left[2\varepsilon_r(i) + \Delta x^2\left(e^{-\psi_i^{(n)}} + e^{\psi_i^{(n)}}\right)\right], \quad c_2(i) = \varepsilon_r(i+1) \tag{11.3}$$

and

$$f^{(n)}(i) = -\Delta x^2\left[\left(e^{-\psi_i^{(n)}} - e^{\psi_i^{(n)}} + C_i\right) + \psi_i^{(n)}\left(e^{-\psi_i^{(n)}} + e^{\psi_i^{(n)}}\right)\right] \tag{11.4}$$

DOI: 10.1201/9781003265849-11

11.2 Poisson Equation in Non-Equilibrium

In non-equilibrium situation we keep densities of electrons and holes which are updated during calculations. They are not explicitly expressed in terms of potential ψ.

In this case Poisson equation is expressed in a condensed form as

$$a_2(i)\psi_{i+1}^{(n+1)} + b_2(i)\psi_i^{(n+1)} + c_2(i)\psi_{i-1}^{(n+1)} = f^{(n)}(i) \qquad (11.5)$$

where

$$a_2(i) = \varepsilon_r(i-1), \quad b_2(i) = -\left[2\varepsilon_r(i) + \Delta x^2 \left(p_i^{(n)} + n_i^{(n)}\right)\right], \quad c_2(i) = \varepsilon_r(i+1) \qquad (11.6)$$

and

$$f^{(n)}(i) = -\Delta x^2 \left[\left(p_i^{(n)} - n_i^{(n)} + C_i\right) + \psi_i^{(n)}\left(p_i^{(n)} + n_i^{(n)}\right)\right] \qquad (11.7)$$

11.3 Electrons

Equation for electrons

$$a_n(i)n_{i-1} + b_n(i)n_i + c_n(i)n_{i+1} = f_{SRH}(i) \qquad (11.8)$$

where

$$a_n(i) = \mu_n(i-\frac{1}{2})B\left(\psi_{i-1} - \psi_i\right) \qquad (11.9)$$

$$b_n(i) = -\left[\mu_n(i-\frac{1}{2})B\left(\psi_i - \psi_{i-1}\right) + \mu_n(i+\frac{1}{2})B\left(\psi_i - \psi_{i+1}\right)\right] \qquad (11.10)$$

$$c_n(i) = \mu_n(i+\frac{1}{2})B\left(\psi_{i+1} - \psi_i\right) \qquad (11.11)$$

11.4 Holes

In a similar way one finds for holes (all variables scaled)

$$a_p(i)p_{i-1} + b_p(i)p_i + c_p(i)p_{i+1} = f_{SRH}(i) \qquad (11.12)$$

where

$$a_p(i) = \mu_p(i-\frac{1}{2})B\left(\psi_i - \psi_{i-1}\right) \qquad (11.13)$$

$$b_p(i) = - \left[\mu_p(i - \frac{1}{2})B\left(\psi_{i-1} - \psi_i\right) + \mu_p(i + \frac{1}{2})B\left(\psi_{i+1} - \psi_i\right) \right] \qquad (11.14)$$

$$c_p(i) = \mu_p(i + \frac{1}{2})B\left(\psi_i - \psi_{i+1}\right) \qquad (11.15)$$

11.5 SRH Recombination

$$f_{SRH}(i) = \Delta x^2 \frac{p_i n_i - 1}{\tau_p\left(n_i + 1\right) + \tau_n\left(p_i + 1\right)} \qquad (11.16)$$

11.6 Currents

Electric current consists of two contributions, electrons and holes. Current due to electrons at the intermediate mesh point $i + \frac{1}{2}$ is (all quantities are not scaled)

$$J_n(i + \frac{1}{2}) = \mu_n(i + \frac{1}{2})q\frac{V_T}{\Delta x}\{n_{i+1}B(\frac{\psi_{i+1} - \psi_i}{V_T}) - n_i B(\frac{\psi_i - \psi_{i+1}}{V_T})\} \quad (11.17)$$

In the program we determine results in a scaled form, therefore current must be expressed in therms of scaled variables. Applying scaling one obtains (we preserved scaled notation)

$$J_n(i + \frac{1}{2}) = q\mu_0 q V_T \frac{C_0}{x_0[cm]}\mu'_n(i + \frac{1}{2})\frac{1}{\Delta x'}\{n'_{i+1}B(\psi'_{i+1} - \psi'_i) - n'_i B(\psi'_i - \psi'_{i+1})\}$$
$$(11.18)$$

The prefactor P becomes (it determines units; we have selected $C_0 = n_{int}$)

$$P = \frac{qV_T n_{int}}{x_0[cm]}\mu_0 = \frac{qn_{int}}{x_0[cm]}D_0, \quad [\frac{C\ cm^{-3}}{cm}\frac{cm^2}{s} = \frac{A}{cm^2}] \qquad (11.19)$$

In the same way one can express current of holes.

11.7 Parameters

As an example we considered structure based on $Al_x Ga_{1-x} As$.

11.7.1 Mobilities

Mobilities for AlGaAs are modelled following Shur [1] as:
  for electrons

$$
\begin{aligned}
\mu_n &= 8000 - 22000x + 10000x^2, \quad \text{(for } x < 0.45) \ [cm^2/Vs] \quad (11.20)\\
&= -255 + 1160x - 720x^2, \quad \text{(for } x > 0.45) \ [cm^2/Vs]
\end{aligned}
$$

for holes
$$
\mu_p = 370 - 970x + 740x^2 \quad [cm^2/Vs] \tag{11.21}
$$

11.7.2 Dielectric constant

Dielectric constant for AlGaAs as a function of Al composition x is

$$
\varepsilon = 13.1 - 3x \tag{11.22}
$$

11.8 Code Summary

For heterostructure analyzed here modified file which creates mesh and also ones which generate data at mesh points. The complete list of files used in this chapter is summarized in Table 11.1. Complete Matlab codes are provided in an Appendix at the end of this chapter.

11.9 Simulated Structures

To test our software we first considered homo-structure fabricated from GaAs.
    The analyzed p-i-n structure based on AlGaAs is shown in Table 11.2. We assumed *Al* composition $x = 0.25$.

11.10 Results

In the following Figures we provide results generated by program in this chapter. They are combined into two groups: results at equilibrium (only generated by Poisson equation) and results at non-equilibrium (obtained using drift-diffusion approach).

TABLE 11.1
List of files used to simulate hetero-diode based on drift-diffusion model.

Listing	Function	Description
Data files		
11.A.1.1	*param_3_layer_AlGaAs.m*	Parameters for AlGaAs
11.A.1.2	*struct_3_layer_pin.m*	Defines simulated structure
Extra		
11.A.2.1	*ber.m*	Implementation of Bernouli function
11.A.2.2	*physical_const.m*	Contains physical constants
11.A.2.3	*scaling_dd.m*	Keeps scaling factors
Models		
11.A.3.1	*mobility_n_AlGaAs_mesh.m*	Mobility of electrons
11.A.3.2	*mobility_p_AlGaAs_mesh.m*	Mobility of holes
11.A.3.3	*recombination_SHR.m*	Recombination model
Main files		
11.A.4.1	*boundary_psi.m*	b.c. for potential
11.A.4.2	*currents.m*	Determines currents
11.A.4.3	*doping_mesh.m*	Assigns doping for all mesh points
11.A.4.4	*electron_density_2.m*	Determines density of electrons
11.A.4.5	*epsilon_mesh.m*	Assigns ε_r for all mesh points
11.A.4.6	*equilibrium_m.m*	Determines quantities in equilibrium
11.A.4.7	*hole_density_2.m*	Determines density of holes
11.A.4.8	*initial.m*	Determines initial potential
11.A.4.9	*mesh_x_CP.m*	Creates 1D mesh
11.A.4.10	*nonequilibrium_m.m*	Determines quantities in nonequilibrium
11.A.4.11	*pin_hetero_CP.m*	Driver
11.A.4.12	*potential_eq_feb.m*	Determines potential in equilibrium
11.A.4.13	*potential_noneq_2.m*	Determines potential in non-equilibrium

11.10.1 Equilibrium case

We plotted potential, electric field and total charge density versus position. They are shown in Figs. 11.1, 11.2 and 11.3.

11.10.2 Non-equilibrium case

The results for non-equilibrium case are shown in Figs. 11.4, 11.5, 11.6 and 11.7.

TABLE 11.2
Simulated structure

Material	Thickness	Doping
AlGaAs	$3\mu m$	$1 \times 10^{17} cm^{-3}$
GaAs	$1\mu m$	$1 \times 10^{15} cm^{-3}$
AlGaAs	$3\mu m$	$1 \times 10^{18} cm^{-3}$

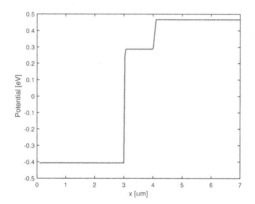

FIGURE 11.1
Potential across the structure in equilibrium case.

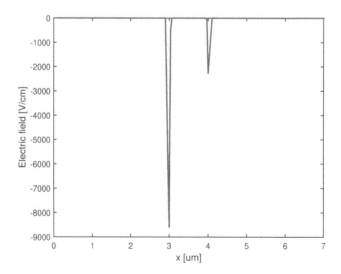

FIGURE 11.2
Electric field across the structure in equilibrium case.

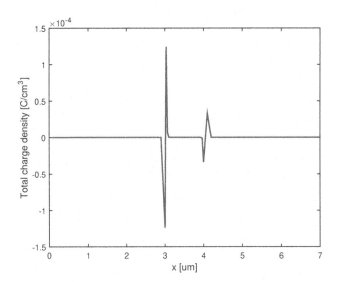

FIGURE 11.3
Total charge across the structure in equilibrium case.

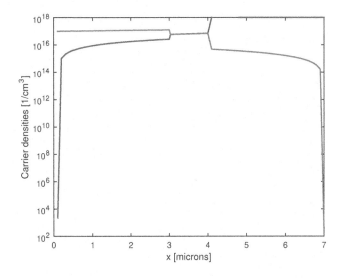

FIGURE 11.4
Densities of electrons and holes across the structure in non-equilibrium case.

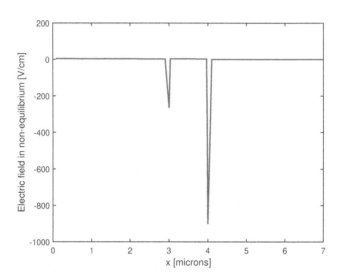

FIGURE 11.5
Electric field across the structure in non-equilibrium case.

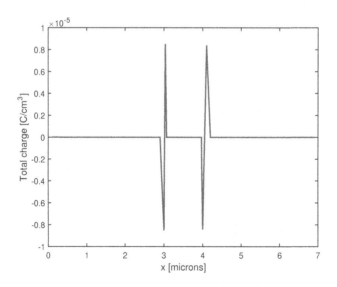

FIGURE 11.6
Total charge across the structure in non-equilibrium case.

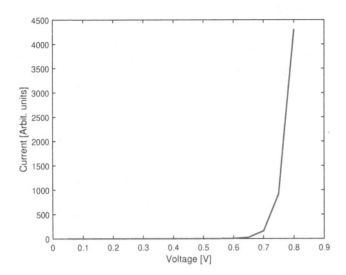

FIGURE 11.7
Current vs voltage in non-equilibrium case.

Appendix 11A: MATLAB Listings

11.10.3   Data files

Listing 11.A.1.1

```
% File name: param_3_layer_AlGaAs.m
%---------------------------------------------------------------
scaling_dd;
%---------------------------------------------------------
% Aluminum composition in different layers
% Al_x Ga_1-x As
%---------------------------------------------------------
x_comp = 0.25;
%
%---------------------------------------------------------
% Values of dielectric constant in each layer
% epsilon = (13.1 - 3x)*eps_zero
%---------------------------------------------------------
eps_layer(3)   = 13.1 - 3*x_comp;
```

```
eps_layer(2)  = 13.1;
eps_layer(1)  = 13.1 - 3*x_comp;
%-------------------------------------------------------------
% Assigns value of mobility for each layer
% Mobility does not depend on electric field
% Simple model of mobilities of electrons for Al_x Ga_(1-x) As
% Based on Shur, p.626

mu_n_layer(3) = 80000-22000*x_comp +10000*x_comp*x_comp;
% [cm^2/(V s]
mu_n_layer(2) = 80000;  % [cm^2/(V s]
mu_n_layer(1) = 80000-22000*x_comp +10000*x_comp*x_comp;
% [cm^2/(V s]
mu_n_layer = mu_n_layer*1d-4;     % convert to [m^2/V s]
mu_n_layer = mu_n_layer/mu_0;            % scaling

mu_p_layer(3) = 370-970*x_comp +740*x_comp*x_comp;
% [cm^2/(V s]
mu_p_layer(2) = 370;  % [cm^2/(V s]
mu_p_layer(1) = 370-970*x_comp +740*x_comp*x_comp;
% [cm^2/(V s]
mu_p_layer = mu_p_layer*1d-4;     % convert to [m^2/V s]
mu_p_layer = mu_p_layer/mu_0;            % scaling


%-------------------------------------------------------------
E_0 = 5;                           % reference energy [eV]
chi_e_GaAs = 4.07;
% Electron affinity for GaAs [eV]
E_g_GaAs = 1.42;                   % Bandgap of GaAs [eV]
chi_e_AlGaAs = 4.07-1.1*x_comp;
% Electron affinity for AlGaAs [eV]
E_g_AlGaAs = 1.424+1.427*x_comp;   % Bandgap of AlGaAs [eV]
band_offset = 0.67;
%-----------------------------
chi_layer(3) = chi_e_GaAs;
chi_layer(2) = chi_e_AlGaAs;
chi_layer(1) = chi_e_GaAs;
E_g_layer(3) = E_g_GaAs ;
E_g_layer(2) = E_g_AlGaAs;
E_g_layer(1) = E_g_GaAs ;
%-------------------------------------------------------------
% Doping values in engineering units
N_A = -1d17;               % acceptor doping [cm^-3]
N_D = 1d18;                % donor doping [cm^-3]
N_neut = 1d15;             % undoped region [cm^-3]
```

```
% Convert to SI system
N_A = N_A*1d6;              % [m^3]
N_D = N_D*1d6;              % [m^3]
N_neut = N_neut*1d6;        % [m^-3]
%-----------------------------------------------------------
dop_layer(3) = N_D;         % in m^-3
dop_layer(2) = N_neut;       % in m^-3
dop_layer(1) = N_A;         % in m^-3
%------------- Scaling of data -------------------------
dop_layer = dop_layer/C_0;
```

Listing 11.A.1.2

```
% File name: struct_3_layer_pin.m
% Input needed to construct mesh
% Contains data.
% Global variables to be transferred to function f_TE.m

global d_layer_int    % thicknesses of internal layers (microns)

% Thicknesses of each layer; values in microns
d_layer(3)   = 3.0;  % microns
d_layer(2)   = 1.0;  % microns
d_layer(1)   = 3.0;  % microns

d_layer_int = d_layer(2);

mesh_layer = [30 30 30];   % number of mesh points in each layer
```

11.10.4   Extra functions

Listing 11.A.2.1

```
% Simplw implementation of Bernouli function

 function b = ber(x)

 if abs(x) >= 1.0d-4
   b = x/(exp(x)-1.0);
 else
   b = 1.0 - 0.5*x*(1.0 + x/6.0)*(1.0-x*x/60.);
 end

 end
```

Listing 11.A.2.2

```
% File name: physical_const.m
% Definitions of basic physical constants

eps_zero   = 8.85d-12;   % Permittivity of free space [F/m]
q      = 1.60d-19;       % elementary charge [C]
k_B    = 1.38d-23;       % Boltzmann constant [J/K]
T      = 300;            % Temperature [K]
n_int    = 1.5d16;       % Intrinsic carrier concentration [m^-3]
```

Listing 11.A.2.3

```
% File name: physical_const.m
% Definitions of basic physical constants

eps_zero   = 8.85d-12;   % Permittivity of free space [F/m]
q      = 1.60d-19;       % elementary charge [C]
k_B    = 1.38d-23;       % Boltzmann constant [J/K]
T      = 300;            % Temperature [K]
n_int    = 1.5d16;       % Intrinsic carrier concentration [m^-3]
```

11.10.5 Models

Listing 11.A.3.1

```
function mu_mesh = mobility_n_AlGaAs_mesh(mesh_layer)
% File name: mobility_n_AlGaAs_mesh.m
% Purpose:   Assigns the value of mobility for AlGaAs for all
% mesh points.
%            That value stays the same within each layer and
% changes
%            from layer to layer.
% mesh_layer)   - number of mesh points in layers
% mu_mesh        - keeps mobility at each mesh point
%
param_3_layer_AlGaAs;

NumberOfLayers = length(eps_layer);
%
i_mesh = 1;
for k = 1:NumberOfLayers              % loop over all layers
    for i = 1:mesh_layer(k)           % loop within layer
        mu_mesh(i_mesh) = mu_n_layer(k);
        i_mesh = i_mesh + 1;
```

```
      end
end
```

Listing 11.A.3.2

```
function mu_mesh = mobility_p_AlGaAs_mesh(mesh_layer)
% File name: mobility_n_AlGaAs_mesh.m
% Purpose:    Assigns the value of mobility for AlGaAs for all
% mesh points.
%             That value stays the same within each layer and
% changes
%             from layer to layer.
% mesh_layer     - mobility of holes within each layer
% mu_mesh        - keeps mobility at each mesh point
%
param_3_layer_AlGaAs;

NumberOfLayers = length(eps_layer);
%
i_mesh = 1;
for k = 1:NumberOfLayers            % loop over all layers
    for i = 1:mesh_layer(k)         % loop within layer
        mu_mesh(i_mesh) = mu_p_layer(k);
        i_mesh = i_mesh + 1;
    end
end
```

Listing 11.A.3.3

```
function fn = recombination_SHR(i,dop,n,p)
% Determines recombination at mesh point 'i'

scaling_dd

tau_n0 = 1e-5;  % SHR recombination time for electrons [seconds]
tau_p0 = 1e-5;  % SHR recombination time for holes [seconds]
N_ref_n = 5e16;
N_ref_p = 5e16;

NN = 2*abs(dop(i));
tau_n = tau_n0/(1 + NN/N_ref_n);    % electron lifetime
tau_p = tau_p0/(1 + NN/N_ref_p);    % hole lifetime

rnum  = n(i)*p(i) - 1;
denom = tau_n*(p(i)+1)+tau_p*(n(i)+1);
```

```
denom = denom/t_0;        % scaling
fn    = (rnum/denom);
end
```

11.10.6   Main files

Listing 11.A.4.1

```
function [psi_L, psi_R] = boundary_psi(N,dop)
% File name: boundary_psi.m
% Provides values of potential at contacts without bias
% Ohmic contacts are assumed
% Contacts are numbered as i=1 (left) and i=N (right) for
% method 2

% left contact
a_1 = 0.5*dop(1);
if (a_1>0)
    temp_1  = a_1*(1.0 + sqrt(1.0+1.0/(a_1*a_1)));
    elseif(a_1<0)
        temp_1 = a_1*(1.0 - sqrt(1.0+1.0/(a_1*a_1)));
end
psi_L = log(temp_1);

% right contact
a_N = 0.5*dop(N);
if (a_N>0)
    temp_N  = a_N*(1.0 + sqrt(1.0+1.0/(a_N*a_N)));
    elseif(a_N<0)
        temp_N = a_N*(1.0 - sqrt(1.0+1.0/(a_N*a_N)));
end
psi_R = log(temp_N);
```

Listing 11.A.4.2

```
function tot_curr_sum = currents(N,mesh_layer,psi,n,p)

% File name: currents.m
% Calculates total current due to electrons and holes

mu_n_mesh = mobility_n_AlGaAs_mesh(mesh_layer);
mu_p_mesh = mobility_p_AlGaAs_mesh(mesh_layer);

tot_curr_sum = 0;
for i = 2:N-1
```

```
            curr_n(i) = mu_n_mesh(i)*(n(i+1)*ber(psi(i+1)
            - psi(i))...
                - n(i)*ber(psi(i) - psi(i+1)));
                % Electron current density

            curr_p(i) = mu_p_mesh(i)*(p(i)*ber(psi(i+1)-psi(i))...
                - p(i+1)*ber(psi(i)-psi(i+1)));
                % Hole current density

        tot_curr(i) = curr_n(i) + curr_p(i);
        tot_curr_sum(i) = tot_curr_sum(i-1)+ tot_curr(i);
end

end
```

Listing 11.A.4.3
```
function dop_mesh = doping_mesh(mesh_layer)
%-----------------------------------------------------------------
% File name: doping_mesh.m
% Purpose:   Assigns the value of doping for all mesh points.
%    That value stays the same within each layer and changes
%            from layer to layer.
% dop_mesh          - keeps doping for each mesh point

param_3_layer_AlGaAs;

NumberOfLayers = length(dop_layer);
%
i_mesh = 1;
for k = 1:NumberOfLayers            % loop over all layers
    for i = 1:mesh_layer(k)         % loop within layer
        dop_mesh(i_mesh) = dop_layer(k);
        i_mesh = i_mesh + 1;
    end
end
```

Listing 11.A.4.4
```
function n = electron_density_2(N,dx,mesh_layer,dop,psi,n,p)
% File name: electron_density_2.m
% Determines electron density using method 2

dx2 = dx*dx;

mu_mesh = mobility_n_AlGaAs_mesh(mesh_layer);
```

```
for i = 2: N-1
        an(i) = mu_mesh(i)*ber(psi(i-1) - psi(i));
        cn(i) = mu_mesh(i)*ber(psi(i+1) - psi(i));
        bn(i) = -(mu_mesh(i)*ber(psi(i) - psi(i-1))
                + mu_mesh(i)*ber(psi(i) - psi(i+1)));
        fn(i) = dx2*recombination_SHR(i,dop,n,p);
end

an(1) = 0;
cn(1) = 0;
bn(1) = 1;
fn(1) = n(1);
an(N) = 0;
cn(N) = 0;
bn(N) = 1;
fn(N) = n(N);

N = N;
for i=1:N
    A(i,i) = bn(i);              % diagonal elements
end
for i=1:N-1
    A(i,i+1) = cn(i);               % above diagonal
end
for i=1:N-1
    A(i+1,i) = an(i+1);              % below diagonal
end

fn=fn';
n = A\fn;                    % 0.329684 seconds. good solution

end
```

Listing 11.A.4.5

```
function eps_mesh = epsilon_mesh(mesh_layer)
%--------------------------------------------------------------------
% File name: epsilon_mesh.m
% Purpose:   Assigns the value of dielectric constant for all
% mesh points.
%            That value stays the same within each layer and
% changes
%            from layer to layer.
```

```
% eps_layer - dielectric constants in each layer
% eps_mesh  - keeps epsilon_dc for each mesh point

param_3_layer_AlGaAs;

NumberOfLayers = length(eps_layer);
%
i_mesh = 1;
for k = 1:NumberOfLayers             % loop over all layers
    for i = 1:mesh_layer(k)          % loop within layer
        eps_mesh(i_mesh) = eps_layer(k);
        i_mesh = i_mesh + 1;
    end
end
```

Listing 11.A.4.6

```
function [psi_eq,n,p] = equilibrium_m(N,x_plot,dx,mesh_layer)
% Function determines equilibrium properties
scaling_dd;
dop = doping_mesh(mesh_layer);       % doping at mesh points
eps_mesh = epsilon_mesh(mesh_layer);

[psi] = initial(N,dop);  % initializes field. No scaling inside

psi_eq = potential_eq_feb(N,dx,psi,dop,eps_mesh);


for i = 2:N-1    % determines equilibrium quantities for plotting
    ro(i) = -n_int*(exp(psi_eq(i))-exp(-psi_eq(i))-dop(i));
    % total charge
    el_field(i) = -(psi_eq(i+1) - psi_eq(i))*V_T/(dx*x_0);
end

for i = 1:N
    n(i) = exp(psi_eq(i));
    p(i) = exp(-psi_eq(i));
end

% Define values at boundaries (for plotting)
el_field(1) = el_field(2);
el_field(N) = el_field(N-1);
ro(1) = ro(2);
ro(N) = ro(N-1);
```

```
toc

% redefine equilibrium quantities for plotting
nf1 = n*n_int*1d-6;            % convert to cm^-3
pf1 = p*n_int*1d-6;            % convert to cm^-3
psi_eq_plot = V_T*psi_eq;      % potential in eV
el_field = el_field*1d-2;      % convert to V/cm
ro_f = q*ro*1d-6;              % convert to C/cm^3

% % %==== Plots of equilibrium values =========
% figure(1)
% plot(x_plot, psi_eq_plot,'LineWidth',1.5)
% xlabel('x [um]');
% ylabel('Potential [eV]');
% pause
%
% figure(2)
% plot(x_plot, el_field,'LineWidth',1.5)
% xlabel('x [um]');
% ylabel('Electric field [V/cm]');
% pause
%
% figure(3)
% plot(x_plot, ro_f,'LineWidth',1.5)
% xlabel('x [um]');
% ylabel('Total charge density [C/cm^3]');
% pause
%
% close all

end
```

Listing 11.A.4.7

```
function p = hole_density_2(N,dx,mesh_layer,dop,psi,n,p)
% File name: hole_density_2.m
% Determines hole density using method 2

dx2 = dx*dx;
mu_mesh = mobility_p_AlGaAs_mesh(mesh_layer);

for i = 2: N-1
    ap(i) =  mu_mesh(i)*ber(psi(i) - psi(i-1));
    cp(i) =  mu_mesh(i)*ber(psi(i) - psi(i+1));
    bp(i) = -(mu_mesh(i)*ber(psi(i-1) - psi(i))
```

```
                 + mu_mesh(i)*ber(psi(i+1) - psi(i)));
        fp(i) = dx2*recombination_SHR(i,dop,n,p);

end

ap(1) = 0;
cp(1) = 0;
bp(1) = 1;
fp(1) = p(1);
ap(N) = 0;
cp(N) = 0;
bp(N) = 1;
fp(N) = p(N);

for i=1:N
    A(i,i) = bp(i);            % diagonal elements
end
for i=1:N-1
    A(i,i+1) = cp(i);          % above diagonal
end
for i=1:N-1
    A(i+1,i) = ap(i+1);        % below diagonal
end

fp=fp';
p = A\fp;

end
```

Listing 11.A.4.8

```
function [psi] = initial(N,dop)
% Initialize potential based on the requirement of charge
% neutrality throughout the whole structure

for i = 1: N
    zz = 0.5*dop(i);
    if(zz > 0)
        xx = zz*(1 + sqrt(1+1/(zz*zz)));
    elseif(zz <  0)
        xx = zz*(1 - sqrt(1+1/(zz*zz)));
    end
    psi(i) = log(xx);
end
```

```
end
```

Listing 11.A.4.9

```
function [N,xn,dx,mesh_layer] = mesh_x_CP()
% Generates one-dimensional mesh along x-axis
% Variable description:
% Output
% N              - number of mesh points
% x              - mesh point coordinates
% dx             - delta x
% mesh_layer     - number of mesh points in each layer
%
struct_3_layer_pin;

NumberOfLayers = length(d_layer);  % determine number of layers
delta = d_layer./mesh_layer;
% separation of points for all layers
%
x(1) = 0.0;                        % coordinate of first mesh point
i_mesh = 1;
for k = 1:NumberOfLayers           % loop over all layers
    for i = 1:mesh_layer(k)        % loop within layer
        x(i_mesh+1) = x(i_mesh) + delta(k);
        i_mesh = i_mesh + 1;
    end
end

N_temp = length(x);
for i=1:N_temp-1
    xn(i) = x(i+1);
end

N = length(xn);        % number of mesh points

dx = sum(d_layer)/N;
% Scaling
```

Listing 11.A.4.10

```
function [el_field_non_eq] = nonequilibrium_m(N,dx,x_plot,
                                 mesh_layer,psi_eq,n,p)

% Solving for the non-equillibirium case

scaling_dd;
```

```
delta_acc = 1E-7;              % Preset the tolerance
dop = doping_mesh(mesh_layer);      % doping at mesh points
eps_mesh = epsilon_mesh(mesh_layer);

Va_max = 0.8;              % max value of applied voltage [V]
dVa = 0.05;               % value of step of applied voltage [V]
% Scale voltages
Va_max = Va_max/V_T;
dVa = dVa/V_T;

psi = psi_eq;

Va  = 0;
Va_max = Va_max;

if(abs(Va_max) == 0)
else
     k = 1;
    while(abs(Va) < abs(Va_max))
        Va = Va + dVa;
        psi(1) = psi(1) + dVa;
        flag = 0;
        k_iter = 0;
        while(~flag)    % convergence loop
            k_iter = k_iter + 1;
            n = electron_density_2(N,dx,mesh_layer,dop,psi,n,p);
            p = hole_density_2(N,dx,mesh_layer,dop,psi,n,p);
            [psi, delta_max] = potential_noneq_2(N,dx,dop,psi,
                                        n,p,eps_mesh);
            if(delta_max < delta_acc)
                flag = 1;
            end
        end            % End of the convergence loop (while loop)
%================================================================

for i = 2:N-1    % determines electric field in non-equilibrium
    el_field_non_eq(i) = -(psi(i+1) - psi(i))*V_T/(dx*x_0);
end
el_field_non_eq(1) = el_field_non_eq(2);
el_field_non_eq(N) = el_field_non_eq(N-1);

% Calculates currents of electrons and holes

tot_curr_sum = currents(N,mesh_layer,psi,n,p);
av_curr(k) = tot_curr_sum(N - 1)*1e-3/(N-2);
```

```
        for i = 2:N-1
            ro_non_eq(i) = - n_int* (n(i) - p(i) - dop(i));
            if(i>1)
            end
        end

        nf1 = n * n_int*1d-6;
        pf1 = p * n_int*1d-6;
        V(k) = Va*V_T;
        k = k + 1;
    end                 % end of voltage loop

aa = q*n_int*D_0/(dx*L_Dint*1d2);    % convert to A/cm^2

av_curr = aa*av_curr;

toc

el_field_non_eq = el_field_non_eq*1d-2;    % convert to V/cm
ro_non_eq = q*ro_non_eq*1d-6;              % convert to C/cm^3

figure(1);
semilogy(x_plot,nf1,'LineWidth',1.5);
% Plotting the final carrier densities
hold on;
semilogy(x_plot,pf1,'LineWidth',1.5);
xlabel('x [microns]');
ylabel('Carrier densities [1/cm^3]');
pause

figure(2);
plot(x_plot,el_field_non_eq,'LineWidth',1.5);
% Plotting the final potential
xlabel('x [microns]');
ylabel('Electric field in non-equilibrium [V/cm]');
pause

figure(3);
plot(x_plot(1:N-1),ro_non_eq,'LineWidth',1.5);
% Plotting the total charge
xlabel('x [microns]');
ylabel('Total charge [C/cm^3]');
pause
```

```
%============== current too large, factor of 10
%===== !!!!!!!!!!!!!!!!!!!!!
figure(4);
plot(V,av_curr,'LineWidth',1.5);
% Plotting the average current
xlabel('Voltage [V]');
ylabel('Current [A/cm^2]');
pause

close all

end
```

Listing 11.A.4.11

```
% File name: pin_hetero_CP.m
% 1D drift-diffusion model of heterogeneous p-n junction
% February, 2023

clear all
close all

tic

'calculating'

scaling_dd;

[N,x_plot,dx,mesh_layer] = mesh_x_CP();
% x in microns, good for plotting

dop = doping_mesh(mesh_layer);       % doping at mesh points
eps_mesh = epsilon_mesh(mesh_layer);

[psi] = initial(N,dop); % initializes field. No scaling inside

psi_eq = potential_eq_feb(N,dx,psi,dop,eps_mesh);


for i = 2:N-1   % determines equilibrium quantities for plotting
    ro(i) = -n_int*(exp(psi_eq(i))-exp(-psi_eq(i))-dop(i));
    % total charge
    el_field(i) = -(psi_eq(i+1) - psi_eq(i))*V_T/(dx*x_0);
end
```

```
for i = 1:N
    n(i) = exp(psi_eq(i));
    p(i) = exp(-psi_eq(i));
end

% Define values at boundaries (for plotting)
el_field(1) = el_field(2);
el_field(N) = el_field(N-1);
ro(1) = ro(2);
ro(N) = ro(N-1);

toc

% redefine equilibrium quantities for plotting
nf1 = n*n_int*1d-6;          % convert to cm^-3
pf1 = p*n_int*1d-6;          % convert to cm^-3
psi_eq_plot = V_T*psi_eq;    % potential in eV
el_field = el_field*1d-2;    % convert to V/cm
ro_f = q*ro*1d-6;            % convert to C/cm^3

% %==== Plots of equilibrium values =========
figure(1)
plot(x_plot, psi_eq_plot,'LineWidth',1.5)
xlabel('x [um]');
ylabel('Potential [eV]');
pause

figure(2)
plot(x_plot, el_field,'LineWidth',1.5)
xlabel('x [um]');
ylabel('Electric field [V/cm]');
pause

figure(3)
plot(x_plot, ro_f,'LineWidth',1.5)
xlabel('x [um]');
ylabel('Total charge density [C/cm^3]');
pause

close all
%-------------------------------------------------------------
% Solving for the non-equillibirium case
delta_acc = 1E-7;            % Preset the tolerance

Va_max = 0.8;                % max value of applied voltage [V]
```

```
dVa = 0.05;                % value of step of applied voltage [V]
% Scale voltages
Va_max = Va_max/V_T;
dVa = dVa/V_T;

psi = psi_eq;

Va  = 0;
Va_max = Va_max;

if(abs(Va_max) == 0)
else
    k = 1;
    while(abs(Va) < abs(Va_max))
        Va = Va + dVa;
        psi(1) = psi(1) + dVa;
        flag = 0;
        k_iter = 0;
        while(~flag)   % convergence loop
            k_iter = k_iter + 1;
            n = electron_density_2(N,dx,mesh_layer,dop,psi,n,p);
            p = hole_density_2(N,dx,mesh_layer,dop,psi,n,p);
            [psi, delta_max] = potential_noneq_2(N,dx,dop,psi,n,
                                        p,eps_mesh);
            if(delta_max < delta_acc)
                flag = 1;
            end
        end               % End of the convergence loop (while loop)
%==============================================================

for i = 2:N-1    % determines electric field in non-equilibrium
    el_field_non_eq(i) = -(psi(i+1) - psi(i))*V_T/(dx*x_0);
end
el_field_non_eq(1) = el_field_non_eq(2);
el_field_non_eq(N) = el_field_non_eq(N-1);

% Calculates currents of electrons and holes

tot_curr_sum = currents(N,mesh_layer,psi,n,p);
av_curr(k) = tot_curr_sum(N - 1)*1e-3/(N-2);

        for i = 2:N-1
            ro_non_eq(i) = - n_int* (n(i) - p(i) - dop(i));
            if(i>1)
            end
```

```
            end

        nf1 = n * n_int*1d-6;
        pf1 = p * n_int*1d-6;
        V(k) = Va*V_T;
        k = k + 1;
    end                     % end of voltage loop

aa = q*n_int*D_0/(dx*L_Dint*1d2);    % convert to A/cm^2

av_curr = aa*av_curr;

toc

el_field_non_eq = el_field_non_eq*1d-2;    % convert to V/cm
ro_non_eq = q*ro_non_eq*1d-6;              % convert to C/cm^3

figure(1);
semilogy(x_plot,nf1,'LineWidth',1.5);
% Plotting the final carrier densities
hold on;
semilogy(x_plot,pf1,'LineWidth',1.5);
xlabel('x [microns]');
ylabel('Carrier densities [1/cm^3]');
pause

figure(2);
plot(x_plot,el_field_non_eq,'LineWidth',1.5);
% Plotting the final potential
xlabel('x [microns]');
ylabel('Electric field in non-equilibrium [V/cm]');
pause

figure(3);
plot(x_plot(1:N-1),ro_non_eq,'LineWidth',1.5);
% Plotting the total charge
xlabel('x [microns]');
ylabel('Total charge [C/cm^3]');
pause

%========== current too large, factor of 10 ===== !!!!!!!!!!!!!!!!!
figure(4);
plot(V,av_curr,'LineWidth',1.5);  % Plotting the average current
xlabel('Voltage [V]');
ylabel('Current [A/cm^2]');
```

```
pause

close all

end
```

Listing 11.A.4.12

```
function psi = potential_eq_feb(N_mesh,dx,psi,dop,ep)
% File name: solution_psi_2.m
% Performs calculations for p-n junction in equilibrium
% using method 2
% Determines psi (potential) in the iteration process using
% Matlab
% functions

physical_const;       % needed for eps_r
dx2 = dx*dx;

error_iter = 0.0001;
relative_error = 1.0;       % Initial value of the relative error

while(relative_error >= error_iter)
    for i = 1: N_mesh
        a(i) = ep(i);
        c(i) = ep(i);
        b(i) = -(2*ep(i)+(dx2)*(exp(-psi(i))+exp(psi(i)))));
        f(i)=(dx2)*(exp(psi(i))-exp(-psi(i))-...
            dop(i)-psi(i)*(exp(psi(i))+exp(-psi(i)))));
        end
% Establishing boundary conditions
a(1) = 0; c(1) = 0; b(1) = 1;
a(N_mesh) = 0; c(N_mesh) = 0; b(N_mesh) = 1;
%
% Establishing boundary conditions for potential at contacts
[psi_L, psi_R] = boundary_psi(N_mesh,dop);

f(1) = psi_L;
f(N_mesh) = psi_R;
%
% Creation of main matrix
for i=1:N_mesh
    A(i,i) = b(i);              % diagonal elements
end
%
for i=1:N_mesh-1
```

```
    A(i,i+1) = c(i);                  % above diagonal
    A(i+1,i) = a(i+1);                % below diagonal
end

ff=f';

psi_1 = A\ff;
psi_1 = psi_1';

delta_psi = psi_1 - psi;
relative_error = max(abs(delta_psi)./psi_1);

psi = psi_1;

end     % end of while loop

end     % end of function
```

Listing 11.A.4.13

```
function [psi, delta_max] = potential_noneq_2(N,dx,dop,psi,n,
    p,ep)
% File name: potential_noneq_2.m
% Solution of the Poisson's equation in non-equibrium
% Using method 2

delta_acc = 1E-7;              % Preset the tolerance

dx2 = dx*dx;
% Define coefficient of main matrix and RHS
for i = 2: N-1
    a(i) = ep(i-1);
    c(i) = ep(i+1);
    b(i) = -(2*ep(i)+dx2*(p(i)+n(i)));
    f(i) = -dx2*(p(i) - n(i) + dop(i) + psi(i)*(p(i) + n(i)));
end

% Initialize values at contacts
a(1) = 0;
c(1) = 0;
b(1) = 1;
f(1) = psi(1);
a(N) = 0;
c(N) = 0;
b(N) = 1;
```

```
f(N) = psi(N);

for i=1:N
    A(i,i) = b(i);              % diagonal elements
end

    A(1,2) = ep(1);
for i=1:N-1
    A(i,i+1) = c(i);              % above diagonal
    A(i+1,i) = a(i+1);           % below diagonal
end

ff=f';

psi_1 = A\ff;
psi_1 = psi_1';

delta = psi_1 - psi;

psi = psi_1;

% Test update in the outer iteration loop

delta_max = 0;
for i = 1: N
    xx = abs(delta(i));
    if(xx > delta_max)
        delta_max=xx;
    end
end

end
```

Bibliography

[1] M. Shur. *Physics of Semiconductor Devices*. Prentice Hall, Englewood Cliffs, NJ, 1990.

12

Multi-Layer Passive Slab Waveguides

In this chapter we will explain how propagation constant is determined and used it to calculate profile of electric field. This part is independent from the rest of the simulator. It is also independently tested. Later-on it will be modified and integrated with the electrical part of the simulator

There are numerous approaches to this problem, (see [1–3] for the summary of some recent results). Here we will develop the simplest one-dimensional model. It is based on work by Kogelnik [4].

This chapter serves as a test of optical part of the simulator. It is not complete, for example we will not include dependency of refractive index on carrier density which is an important effect. It will be included later on. Also, we will not scale initial equations, which will be done in a later chapter. We will mostly concentrate on one-dimensional situation and TE mode.

12.1 Modes of the Arbitrary Three Layer Asymmetric Planar Waveguide in 1D

Before we start testing our approach and software for several structures, lets try to get some intuition about the existence of possible modes and profiles of electric field for the simplest structure which consists of only three layers. Such structure can be easy analyzed analytically.

Consider the following structure, as shown in Fig. 12.1. Here, $\overline{n}_c$ signifies refractive index of cladding, $\overline{n}_f$ of film, and $\overline{n}_s$ that of substrate. For asymmetric slab, $\overline{n}_c \neq \overline{n}_s$.

Define the following quantities

$$
\begin{aligned}
\kappa_c^2 &= \overline{n}_c^2 k^2 - \beta^2 \equiv -\gamma_c^2 \\
\kappa_f^2 &= \overline{n}_f^2 k^2 - \beta^2 \\
\kappa_s^2 &= \overline{n}_s^2 k^2 - \beta^2 \equiv -\gamma_s^2
\end{aligned}
\tag{12.1}
$$

DOI: 10.1201/9781003265849-12

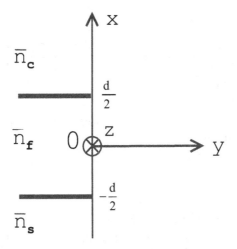

FIGURE 12.1
Waveguiding structure (3-layers) used for TE-mode analysis.

where γ_i — transverse decay and κ_i — propagation constants. i takes the values c, f, s as appropriate. In the following we discuss only TE modes for this 3-layer structure.

For guided TE modes the following solutions exist

$$
\begin{array}{llll}
E_y(x) = A_c e^{-\gamma_c(x-d/2)} & \frac{d}{2} < x & \text{(cover)} \\
E_y(x) = A\cos\kappa_f x + B\sin\kappa_f x & -\frac{d}{2} < x < \frac{d}{2} & \text{(film)} & (12.2) \\
E_y(x) = A_s e^{\gamma_s(x+d/2)} & x < -\frac{d}{2} & \text{(substrate)}
\end{array}
$$

Calculate derivatives

$$
\begin{array}{llll}
\frac{dE_y(x)}{dx} = -\gamma_c A_c e^{-\gamma_c(x-d/2)} & \frac{d}{2} < x & \text{(cover)} \\
\frac{dE_y(x)}{dx} = -\kappa_f A\cos\kappa_f x + \kappa_f B\sin\kappa_f x & -\frac{d}{2} < x < \frac{d}{2} & \text{(film)} & (12.3) \\
\frac{dE_y(x)}{dx} = \gamma_s A_s e^{\gamma_s(x+d/2)} & x < -\frac{d}{2} & \text{(substrate)}
\end{array}
$$

Boundary conditions

$$
E_y \text{ and } \frac{dE_y(x)}{dx} \text{ are continuous for } x = \frac{d}{2} \text{ and for } x = -\frac{d}{2} \qquad (12.4)
$$

Applying boundary conditions for E_y and $\frac{dE_y(x)}{dx}$ results in the following equations

$$
\begin{array}{ll}
\text{for } x = -\frac{d}{2} & A\cos\kappa_f\frac{d}{2} - B\sin\kappa_f\frac{d}{2} = A_s \\
& \kappa_f A\sin\kappa_f\frac{d}{2} + \kappa_f B\cos\kappa_f\frac{d}{2} = \gamma_s A_s
\end{array} \qquad (12.5)
$$

$$
\begin{array}{ll}
\text{for } x = \frac{d}{2} & A_c = A\cos\kappa_f\frac{d}{2} + B\sin\kappa_f\frac{d}{2} \\
& -\gamma_c A_c = -\kappa_f A\sin\kappa_f\frac{d}{2} + \kappa_f B\cos\kappa_f\frac{d}{2}
\end{array}
$$

The above equations can be written in a matrix form

$$
\begin{bmatrix}
\cos \kappa_f \frac{d}{2} & -\sin \kappa_f \frac{d}{2} & -1 & 0 \\
\kappa_f \sin \kappa_f \frac{d}{2} & \kappa_f \cos \kappa_f \frac{d}{2} & -\gamma_s & 0 \\
\cos \kappa_f \frac{d}{2} & \sin \kappa_f \frac{d}{2} & 0 & -1 \\
-\kappa_f \sin \kappa_f \frac{d}{2} & \kappa_f \cos \kappa_f \frac{d}{2} & 0 & \gamma_c
\end{bmatrix}
\begin{bmatrix}
A \\
B \\
A_s \\
A_c
\end{bmatrix} = 0 \qquad (12.6)
$$

For the above homogeneous system to have nontrivial solution, the main determinant should vanish

$$
\begin{vmatrix}
\cos \kappa_f \frac{d}{2} & -\sin \kappa_f \frac{d}{2} & -1 & 0 \\
\kappa_f \sin \kappa_f \frac{d}{2} & \kappa_f \cos \kappa_f \frac{d}{2} & -\gamma_s & 0 \\
\cos \kappa_f \frac{d}{2} & \sin \kappa_f \frac{d}{2} & 0 & -1 \\
-\kappa_f \sin \kappa_f \frac{d}{2} & \kappa_f \cos \kappa_f \frac{d}{2} & 0 & \gamma_c
\end{vmatrix} = 0 \qquad (12.7)
$$

The above determinant is evaluated as follows. Expand it over last column and obtain

$$
\gamma_c
\begin{vmatrix}
\cos \kappa_f \frac{d}{2} & -\sin \kappa_f \frac{d}{2} & -1 \\
\kappa_f \sin \kappa_f \frac{d}{2} & \kappa_f \cos \kappa_f \frac{d}{2} & -\gamma_s \\
\cos \kappa_f \frac{d}{2} & \sin \kappa_f \frac{d}{2} & 0
\end{vmatrix}
+
\begin{vmatrix}
\cos \kappa_f \frac{d}{2} & -\sin \kappa_f \frac{d}{2} & -1 \\
\kappa_f \sin \kappa_f \frac{d}{2} & \kappa_f \cos \kappa_f \frac{d}{2} & -\gamma_c \\
-\kappa_f \sin \kappa_f \frac{d}{2} & \kappa_f \cos \kappa_f \frac{d}{2} & 0
\end{vmatrix} = 0
$$

Evaluating both determinants we obtain

$$
\sin^2 \kappa_f \frac{d}{2} - \cos^2 \kappa_f \frac{d}{2} + \frac{\kappa_f}{\gamma_c} \sin \kappa_f \frac{d}{2} \cos \kappa_f \frac{d}{2} - \frac{\gamma_s}{\kappa_f} \sin \kappa_f \frac{d}{2} \cos \kappa_f \frac{d}{2} = 0
$$

which can be expressed as

$$
\tan^2 \kappa_f \frac{d}{2} - 1 + \frac{\kappa_f}{\gamma_c} \tan \kappa_f \frac{d}{2} - \frac{\gamma_s}{\kappa_f} \tan \kappa_f \frac{d}{2} = 0 \qquad (12.8)
$$

The above is the general equation for 3-layer asymmetric waveguide. For symmetric waveguide

$$
\gamma_s = \gamma_c = \gamma
$$

one can write Eq.12.8 as

$$
\left(\tan \kappa_f \frac{d}{2} - \frac{\gamma}{\kappa_f} \right) \left(\tan \kappa_f \frac{d}{2} + \frac{\kappa_f}{\gamma} \right) = 0 \qquad (12.9)
$$

Solving numerically the above transcendental equation one obtains propagation constant β which enters κ_f and γ via definitions (12.1). Different types of modes for various values of propagation constants are shown in Fig. 12.2 (adopted from [5]). Profiles of several modes are shown.

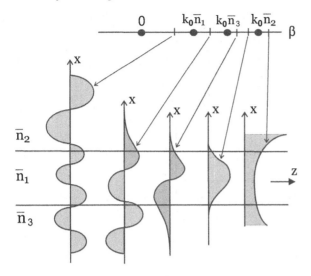

FIGURE 12.2
Different types of modal solutions for three layer structure.

Note: The above modes are not normalized for power. Power P carried by a mode per unit quide width is as follows

$$
\begin{aligned}
P &= -2 \int_{-\infty}^{+\infty} dx\, E_y H_x \\
&= \frac{2\beta}{\omega\mu} \int_{-\infty}^{+\infty} dx\, E_y^2 \\
&= N \sqrt{\frac{\varepsilon_0}{\mu_0}} E_f^2 \cdot h_{eff} \\
&= E_f \cdot H_f \cdot h_{eff}
\end{aligned}
$$

where $h_{eff} \equiv h + \frac{1}{\gamma_s} + \frac{1}{\gamma_c}$ is the effective thickness of the waveguide.

12.2 Multilayer Waveguide

In this section we develop propagation matrix formalism and use it to analyze multilayered waveguiding structures. We follow our work [6].

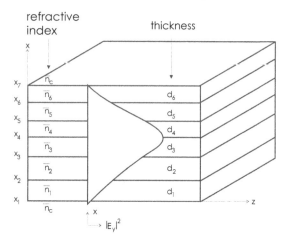

FIGURE 12.3
Multilayer waveguide.

12.2.1   Propagation matrix formulation

Consider general multilayer optical waveguiding structure (Fig. 12.3). Cartesian coordinate system is also shown with x-axis being perpendicular to the waveguide interface. Light will be thus confined in the x-direction. Usually, the top layer creates cladding region and the lower layer is a substrate. In both, cladding and substrate field decays exponentially. Typical field profile is also shown in Fig. 12.3 . The refractive index $\bar{n}(x)$ and the corresponding electric field will be functions of only x-coordinate. The y-dependence is neglected by setting $\partial/\partial y = 0$ which corresponds to the assumption that the waveguide is infinitely extended in y-direction. The structure depicted in Fig. 12.3 consists in general of N layers with arbitrary $i - th$ layer having thickness d_i and and (complex) refractive index $\bar{n}_i(x) = \overline{n}_{real,i}(x) - j\alpha_i(x)$. First, lets summarize equations derived earlier (in Chapter 1) for one-dimensional situation and TE mode (where $E_x = E_z = H_y = 0$). Assuming harmonic fields and introducing propagation constant β in the usual way, one obtains the following equations

$$\frac{dE_y}{dx} = -j\omega\mu_0 H_z \tag{12.10}$$

$$-j\beta E_y = j\omega\mu_0 H_x \tag{12.11}$$

$$\frac{dH_z}{dx} + j\beta H_x = -j\omega\varepsilon_0\varepsilon_r E_y \tag{12.12}$$

The above system can be combined together to obtain wave equation. For $i - th$ layer it reads

$$\frac{d^2 E_{yi}(x)}{dx^2} = \left\{ \beta^2 - k_0^2 \bar{n}_i^2(x) \right\} E_{yi}(x) \tag{12.13}$$

where $k_0 = \omega\sqrt{\mu_0\varepsilon_0}$ and $\varepsilon_r = \overline{n}_i^2$. $\overline{n}_i$ is known as refractive index for layer $i - th$ layer. At this stage it is customary to define

$$\kappa_i^2 = k_0^2\overline{n}_i^2 - \beta^2 \tag{12.14}$$

The solution to the wave equation (12.13) in $i - th$ layer is

$$E_{yi}(x) = A_i e^{-j\kappa_i(x-x_i)} + B_i e^{j\kappa_i(x-x_i)} \tag{12.15}$$

At this stage the following notation is customary introduced

$$U_i(x) = E_{yi}(x) \tag{12.16}$$
$$V_i(x) = \omega\mu_0 H_{zi}(x) \tag{12.17}$$

From Eq.(12.10) one obtains for a new variable $U_i(x)$

$$\frac{dU_i(x)}{dx} = -jV_i(x) \tag{12.18}$$

Similarly, from (12.11) and (12.12) one finds equation for $V_i(x)$. First, eliminate H_x in Eq. (12.12)

$$\frac{dH_{zi}}{dx} + j\beta\frac{(-\beta)}{\omega\mu_0}E_{yi} = -j\omega\varepsilon_0\varepsilon_r E_{yi}$$

Second, multiply both sides of the last equation by $\omega\mu_0$

$$\omega\mu_0\frac{dH_{zi}}{dx} - j\beta^2 E_{yi} = -j\omega^2\mu_0\varepsilon_0\varepsilon_r E_{yi}$$
$$= -jk_0^2\overline{n}_i^2 E_{yi}$$

Using definitions gives

$$\frac{dV_i(x)}{dx} = -j\kappa_i^2 U_i(x) \tag{12.19}$$

Equations (12.18) and (12.19) can be written in matrix form [4, 7]

$$\frac{d}{dx}\begin{pmatrix} U_i(x) \\ V_i(x) \end{pmatrix} = \begin{pmatrix} 0 & -j \\ -j\kappa_i^2 & 0 \end{pmatrix}\begin{pmatrix} U_i(x) \\ V_i(x) \end{pmatrix} \tag{12.20}$$

The solutions of this equation for $i - th$ layer are given by

$$U_i(x) = A_i e^{-j\kappa_i(x-x_i)} + B_i e^{j\kappa_i(x-x_i)} \tag{12.21}$$

$$V_i(x) = j\frac{dU_i(x)}{dx} = \kappa_i\{A_i e^{-j\kappa_i(x-x_i)} - B_i e^{j\kappa_i(x-x_i)}\} \tag{12.22}$$

The above solutions can be written in a compact matrix form

$$\left[\begin{array}{c} U_i(x) \\ V_i(x) \end{array} \right] = \left[\begin{array}{cc} e^{-j\kappa_i(x-x_i)} & e^{j\kappa_i(x-x_i)} \\ \kappa_i e^{-j\kappa_i(x-x_i)} & -\kappa_i e^{j\kappa_i(x-x_i)} \end{array} \right] \left[\begin{array}{c} A_i \\ B_i \end{array} \right] \qquad (12.23)$$

From the above, at $x = x_i$ one finds

$$\left[\begin{array}{c} U_i(x_i) \\ V_i(x_i) \end{array} \right] = \left[\begin{array}{cc} 1 & 1 \\ \kappa_i & -\kappa_i \end{array} \right] \left[\begin{array}{c} A_i \\ B_i \end{array} \right] \qquad (12.24)$$

Inverting last equation gives the expression for coefficients in terms of field and its derivative

$$\left[\begin{array}{c} A_i \\ B_i \end{array} \right] = \frac{1}{2} \left[\begin{array}{cc} 1 & \frac{1}{j\kappa_i} \\ 1 & -\frac{1}{j\kappa_i} \end{array} \right] \left[\begin{array}{c} U_i(x) \\ V_i(x) \end{array} \right] \qquad (12.25)$$

Substituting the last expression into Eq.(12.23) and performing the required matrix multiplication results in propagation matrix formula

$$\left[\begin{array}{c} U_i(x) \\ V_i(x) \end{array} \right] = T_i(x_i) \left[\begin{array}{c} U_i(x_i) \\ V_i(x_i) \end{array} \right] \qquad (12.26)$$

where the propagation matrix $T_i(x_i)$ is

$$T_i(x_i) = \left[\begin{array}{cc} \cos\kappa_i(x-x_i) & -\frac{j}{\kappa_i}\sin\kappa_i(x-x_i) \\ -j\kappa_i\sin\kappa_i(x-x_i) & \cos\kappa_i(x-x_i) \end{array} \right] \qquad (12.27)$$

Equation (12.26) describes propagation of fields E_{yi} and H_{zi} from point x_i to an arbitrary point x within $i-th$ layer.

12.2.2 Propagation constant

Propagation constant β is obtained if we propagate fields across all interfaces from substrate to cladding. First, consider propagation between two consequitive interfaces x_i and x_{i+1}. From Eqs. (12.26) and (12.27) one finds

$$\left[\begin{array}{c} U_{i+1}(x_{i+1}) \\ V_{i+1}(x_{i+1}) \end{array} \right] = \left[\begin{array}{cc} \cos\kappa_i(x_{i+1}-x_i) & -\frac{j}{\kappa_i}\sin\kappa_i(x_{i+1}-x_i) \\ -j\kappa_i\sin\kappa_i(x_{i+1}-x_i) & \cos\kappa_i(x_{i+1}-x_i) \end{array} \right] \left[\begin{array}{c} U_i(x_i) \\ V_i(x_i) \end{array} \right]$$
$$(12.28)$$

From practical point it is more convenient to use layer thickness then location of layers. Lets then introduce

$$d_i = x_{i+1} - x_i \qquad (12.29)$$

Propagation matrix which appears in Eq.(12.28) and which propagates field from x_i to x_{i+1} will now be called M_i and it has the form

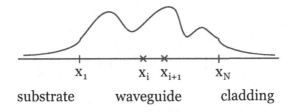

FIGURE 12.4
Profile of electromagnetic field in a 1D waveguide.

$$M_i = \begin{bmatrix} \cos \kappa_i d_i & -\frac{j}{\kappa_i} \sin \kappa_i d_i \\ -j\kappa_i \sin \kappa_i d_i & \cos \kappa_i d_i \end{bmatrix} \qquad (12.30)$$

The above propagation matrix propagates electric field (and its derivative) in the direction from right to left, i.e. from point x_i to point x_{i+1}. That formalism can now be used to propagate electric field through guiding and active regions, from point on the right (substrate) to point on the left (cladding), as

$$\begin{bmatrix} U_c \\ V_c \end{bmatrix} = \prod_{i=1}^{N} M_i \begin{bmatrix} U_s \\ V_s \end{bmatrix}$$

$$= \begin{bmatrix} m_{11} & m_{12} \\ m_{21} & m_{22} \end{bmatrix} \begin{bmatrix} U_s \\ V_s \end{bmatrix} \qquad (12.31)$$

Here N is the number of layers. In the substrate and cladding the fields decay exponentially. Their corresponding expressions for fields in substrate and cladding are found from (12.21) and (12.22) and are (see Fig. 12.4):
for $x < x_1$

$$U_s(x) = A_s e^{\gamma_s (x-x_1)} \qquad (12.32)$$
$$V_s(x) = j\gamma_s A_s e^{\gamma_s (x-x_1)} \qquad (12.33)$$

for $x > x_N$

$$U_c(x) = A_c e^{-\gamma_c (x-x_N)} \qquad (12.34)$$
$$V_c(x) = -j\gamma_c A_c e^{-\gamma_c (x-x_N)} \qquad (12.35)$$

For the existence of propagating modes, fields have arbitrary behavior inside waveguide and decay exponentially within substrate and cladding. In the above, $\gamma_i^2 = \beta^2 - k_0^2 \bar{n}_i^2$. In the above the following relation has been used $\gamma_i = j\kappa_i$. From Eq.(12.31) one obtains the following equations which connect cladding and substrate regions

$$U_c = m_{11}U_s + m_{12}V_s \qquad (12.36)$$
$$V_c = m_{21}U_s + m_{22}V_s \qquad (12.37)$$

Substituting expressions (12.32) and (12.34) into above equations, one finally obtains the required dispersion relation

$$\gamma_c\gamma_s m_{12} - m_{21} = j(\gamma_c m_{11} + \gamma_s m_{22}) \qquad (12.38)$$

The last expression (12.38) are used to find propagation constant β. Those steps must be carried-out numerically.

12.2.3   Electric field

Once propagation constant is found, one can obtain profile of electric field corresponding to that propagation constant. To do this, let's write Eq.(12.26) explicitly

$$\begin{bmatrix} U_i(x) \\ V_i(x) \end{bmatrix} = \begin{bmatrix} \cos\kappa_i(x-x_i) & -\frac{j}{\kappa_i}\sin\kappa_i(x-x_i) \\ -j\kappa_i\sin\kappa_i(x-x_i) & \cos\kappa_i(x-x_i) \end{bmatrix} \begin{bmatrix} U_i(x_i) \\ V_i(x_i) \end{bmatrix}$$
$$(12.39)$$

where $x_i < x < x_{i+1}$ and $\kappa_i^2 = k_0^2\overline{n}_i^2 - \beta^2$. The field amplitude at a reference layer, say x_1 can be chosen arbitrarily. Choosing a value for U_s at x_1 in Eq. (12.32), sets also the value of V_s at the same point. Equation (12.39) is then employed to determine fields at other points of the structure in terms of initial arbitrary value of U_s.

12.3   Testing

We tested two structures one of which will be later considered as laser.

12.3.1   6-layer lossy waveguide.

First structure tested by our program is defined in Table 12.1, from [8]. It is a 6-layer lossy waveguide structure operating at the wavelength $\lambda = 1.523\mu m$. Propagation constants for this structure are summarized in Table 12.2. Obtained results show an excellent agreement with the published data.

The TE electric field distribution of the fundamental mode for this structure is shown in Fig. 12.5.

TABLE 12.1
Geometry of a six-layer lossy dielectric waveguide.

Layer	Refractive index	Thickness	Description
D_8	$\bar{n}_8 = 1.00$	$1\mu m$	Cladding
D_7	$\bar{n}_7 = 3.38327$	$0.10\mu m$	
D_6	$\bar{n}_6 = 3.39614$	$0.20\mu m$	
D_5	$\bar{n}_5 = 3.5321 - j0.08817$	$0.60\mu m$	
D_4	$\bar{n}_4 = 3.39583$	$0.518\mu m$	
D_3	$\bar{n}_3 = 3.22534$	$1.60\mu m$	
D_2	$\bar{n}_2 = 3.16455$	$0.60\mu m$	
D_1	$\bar{n}_1 = 3.172951$	$1\mu m$	Substrate

TABLE 12.2
Results of propagation constants for six-layer lossy
dielectric waveguide.

Mode	Propagation constants
TE_0	3.460829693510364 + 0.072663342917385i
TE_1	3.316707802046375 + 0.023275817588121i
TE_2	3.208555428734457 + 0.012782067986633i
TE_3	3.195490593396514 + 0.012585955654404i

12.3.2 p-i-n structure

This structure is summarizing in Table 12.3. It consists of only 3-layers: $p - i - n$, where i indicates undoped region which practically means much smaller than p or n doped regions. Assumed Al composition $x = 0.25$ and operating wavelength $\lambda = 1.3\mu m$.

Propagation constant of the fundamental TE mode of this structure is 3.564449466339109. Electric field profile of this mode is shown in Fig. 12.6.

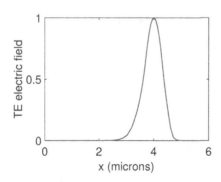

FIGURE 12.5
Field profile of fundamental TE mode for a six-layer lossy waveguide.

TABLE 12.3

Geometry of p-i-n structure.

Layer	Thickness	Mesh points	Doping
3	$3\mu m$	30	$1 \times 10^{17} cm^{-3}$
2	$0.5\mu m$	30	$1 \times 10^{15} cm^{-3}$
1	$3\mu m$	30	$1 \times 10^{18} cm^{-3}$

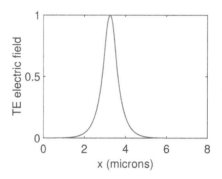

FIGURE 12.6

Field profile of fundamental TE mode for p-i-n structure.

12.4   List of Files

List of files is provided in Table 12.4.

Appendix 12A: MATLAB Listings

12.4.1   Data files

Listing 12.A.1.1

Contains parameters for p-i-n structure.

```
% File name: param_3_layer_AlGaAs.m
%----------------------------------------------------------------
scaling_dd;
%----------------------------------------------------------------
% Aluminum composition in different layers
% Al_x Ga_1-x As
%----------------------------------------------------------------
```

TABLE 12.4
List of Matlab functions for Chapter 12. Analysis of slab waveguide.

Listing	Function	Description
Data files		
12.A.1.1	*param_3_layer_AlGaAs.m*	parameters used pin structure
12.A.1.2	*param_lossy.m*	parameters used in in [7]
12.A.1.3	*struct_3_layer_pin.m*	pin structure
12.A.1.4	*struct_lossy.m*	lossy structure
General		
12.A.2.1	*optical_global.m*	transfer of global variables and parameters
12.A.2.2	*parameters.m*	control of used parameters
12.A.2.3	*physical_const.m*	contains physical constants
12.A.2.4	*scaling_dd.m*	contains scaling factors
12.A.2.5	*structure.m*	controls used structure
Main files		
12.A.3.1	*epsilon_mesh.m*	assigns epsilon values
12.A.3.2	*f_TE.m*	use to determine electric field
12.A.3.3	*mesh_x_CP.m*	creates mesh
12.A.3.4	*muller.m*	implements Muller method
12.A.3.5	*slab.m*	Driver
12.A.3.6	*TE_field.m*	solves for TE electric field

```
x_comp = 0.25;
%
%----------------------------------------------------------
% Values of dielectric constant in each layer
% epsilon = (13.1 - 3x)*eps_zero
%----------------------------------------------------------
eps_layer(3)  = 13.1 - 3*x_comp;
eps_layer(2)  = 13.1;
eps_layer(1)  = 13.1 - 3*x_comp;
%----------------------------------------------------------
% Values of refractive index needed for optical calculations
n_layer = sqrt(eps_layer);
n_layer_int = n_layer(2);
n_c = n_layer(3);
n_s = n_layer(1);

%----------------------------------------------------------
% Assigns value of mobility for each layer
% Mobility does not depend on electric field
% Simple model of mobilities of electrons for Al_x Ga_(1-x) As
% Based on Shur, p.626

mu_n_layer(3) = 80000-22000*x_comp +10000*x_comp*x_comp;
```

```
% [cm^2/(V s)]
mu_n_layer(2) = 80000;  % [cm^2/(V s)]
mu_n_layer(1) = 80000-22000*x_comp +10000*x_comp*x_comp;
% [cm^2/(V s)]
mu_n_layer = mu_n_layer*1d-4;       % convert to [m^2/V s]
mu_n_layer = mu_n_layer/mu_0;               % scaling

mu_p_layer(3) = 370-970*x_comp +740*x_comp*x_comp;  % [cm^2/(V s)]
mu_p_layer(2) = 370;  % [cm^2/(V s)]
mu_p_layer(1) = 370-970*x_comp +740*x_comp*x_comp;  % [cm^2/(V s)]
mu_p_layer = mu_p_layer*1d-4;       % convert to [m^2/V s]
mu_p_layer = mu_p_layer/mu_0;               % scaling

%------------------------------------------------------------
E_0 = 5;                               % reference energy [eV]
chi_e_GaAs = 4.07;
% Electron affinity for GaAs [eV]
E_g_GaAs = 1.42;                       % Bandgap of GaAs [eV]
chi_e_AlGaAs = 4.07-1.1*x_comp;
% Electron affinity for AlGaAs [eV]
E_g_AlGaAs = 1.424+1.427*x_comp;    % Bandgap of AlGaAs [eV]
band_offset = 0.67;
%----------------------------
chi_layer(3) = chi_e_GaAs;
chi_layer(2) = chi_e_AlGaAs;
chi_layer(1) = chi_e_GaAs;
E_g_layer(3) = E_g_GaAs ;
E_g_layer(2) = E_g_AlGaAs;
E_g_layer(1) = E_g_GaAs ;
%------------------------------------------------------------
% Doping values in engineering units
N_A = -1d17;                   % acceptor doping [cm^-3]
N_D = 1d18;                    % donor doping [cm^-3]
N_neut = 1d15;                 % undoped region [cm^-3]
% Convert to SI system
N_A = N_A*1d6;                 % [m^3]
N_D = N_D*1d6;                 % [m^3]
N_neut = N_neut*1d6;           % [m^-3]
%------------------------------------------------------------
dop_layer(3) = N_D;            % in m^-3
dop_layer(2) = N_neut;          % in m^-3
dop_layer(1) = N_A;            % in m^-3
%------------- Scaling of data ------------------------
dop_layer = dop_layer/C_0;
```

Listing 12.A.1.2
Contains parameters for lossy structure.

```
% File name: param_lossy.m
% Contains data for lossy waveguide.
% Reference:
% C. Chen et al, Proc. SPIE, v.3795 (1999)

n_c              = 1.0;         % cladding
n_s              = 3.172951;    % substrate

n_layer_int(6) = 3.38327;
n_layer_int(5) = 3.39614;
n_layer_int(4) = 3.5321-1j*0.08817;
n_layer_int(3) = 3.39583;
n_layer_int(2) = 3.22534;
n_layer_int(1) = 3.16455;

eps_layer(8) = n_c^2;
eps_layer(7) = n_layer_int(6)^2;
eps_layer(6) = n_layer_int(5)^2;
eps_layer(5) = n_layer_int(4)^2;
eps_layer(4) = n_layer_int(3)^2;
eps_layer(3) = n_layer_int(2)^2;
eps_layer(2) = n_layer_int(1)^2;
eps_layer(1) = n_s^2;
```

Listing 12.A.1.3
Defines p-i-n structure.

```
% File name: struct_3_layer_pin.m
% Input needed to construct mesh
% Contains data.
% Global variables to be transferred to function f_TE.m

% Thicknesses of each layer; values in microns
d_layer(3)  = 3.0;  % microns
d_layer(2)  = 0.5;  % microns
d_layer(1)  = 3.0;  % microns

d_layer_int = d_layer(2);   % thickness of internal layer

mesh_layer = [30 30 30];  % number of mesh points in each layer

lambda     = 1.3;  % wavelength in microns
```

```
k_0        = 2*pi/lambda;
```

Listing 12.A.1.4
Defines lossy waveguiding structure.

```
% File name: struct_lossy.m
% Input needed to construct mesh

% Thickness of each layer; values in microns
d_layer(8) = 1.0;    % cladding
d_layer(7) = 0.1;
d_layer(6) = 0.2;
d_layer(5) = 0.6;
d_layer(4) = 0.518;
d_layer(3) = 1.6;
d_layer(2) = 0.6;
d_layer(1) = 1.0;    % substrate

d_layer_int(6) = 0.1;
d_layer_int(5) = 0.2;
d_layer_int(4) = 0.6;
d_layer_int(3) = 0.518;
d_layer_int(2) = 1.6;
d_layer_int(1) = 0.6;

mesh_layer = [10 10 10 10 10 10 10 10];
lambda     = 1.523;  % wavelength in microns
k_0        = 2*pi/lambda;
```

12.4.2   Extra files

Listing 12.A.2.1
Controls transfer of global variables and parameters.

```
% File name: optical_global.m
% Contains variables needed in optical calculations
% Global variables to be transferred to function f_TE.m
% Contains data for 1D waveguide.

global n_c          % ref. index cladding
global n_layer_int    % ref. index of internal layers
global n_s          % ref. index substrate
%global d_c          % thickness of cladding (microns)
global d_layer_int    % thicknesses of internal layers (microns)
```

```
%global d_s          % thickness of substrate (microns)
global k_0           % wavenumber
%
% Here simulated structure is defined
structure;
parameters;
```

Listing 12.A.2.2
Function controls which parameters are used.

```
% File name: parameters.m
% Selection of parameters is done here

param_3_layer_AlGaAs;
%param_lossy;
```

Listing 12.A.2.3
Function contains physical constants.

```
% File name: physical_const.m
% Definitions of basic physical constants

eps_zero  = 8.85d-12;  % Permittivity of free space [F/m]
q      = 1.60d-19;     % elementary charge [C]
k_B    = 1.38d-23;     % Boltzmann constant [J/K]
T      = 300;          % Temperature [K]
n_int   = 1.5d16;      % Intrinsic carrier concentration [m^-3]
```

Listing 12.A.2.4
Function contains scaling factors.

```
% File name: scaling_dd.m
% Purpose: keeps general and electrical scaling factors
% for dd model
% Variable description:
% Basic scaling:
% V_T     - thermal voltage
% x_0  - length scaling
% C_0  - concentration scaling
% D_0  - diffusion scaling
%
% Derived scaling:
% t_0  - time scaling
% mu_0 - mobility scaling
```

```
physical_const;          % needed for V_T, n_int

V_T = k_B*T/q;           % Thermal voltage[eV]
L_Dint = sqrt(eps_zero*V_T/(q*n_int));

x_0 = L_Dint;            % unit [meters]
C_0 = n_int;             % unit [m^-3]
D_0 = 1d3;            % [cm^2/s]
t_0 = (x_0^2)/D_0;
mu_0 = D_0/V_T;
```

Listing 12.A.2.5
Function controls which structure is simulated.

```
% File name: structure.m
% Selection of the appropriate structure is done here

struct_3_layer_pin;
%struct_lossy;
```

12.4.3   Main files

Listing 12.A.3.1
Assigns values of epsilon for each mesh point.

```
function eps_mesh = epsilon_mesh(mesh_layer)
%-------------------------------------------------------------------
% File name: epsilon_mesh.m
% Purpose:   Assigns the value of dielectric constant for all
% mesh points.
%            That value stays the same within each layer and
% changes
%            from layer to layer.
% eps_mesh       - keeps epsilon_dc for each mesh point

%param_3_layer_AlGaAs;
parameters;

NumberOfLayers = length(eps_layer);
%
i_mesh = 1;
for k = 1:NumberOfLayers          % loop over all layers
    for i = 1:mesh_layer(k)          % loop within layer
```

```
        eps_mesh(i_mesh) = eps_layer(k);
        i_mesh = i_mesh + 1;
    end
end
```

Listing 12.A.3.2
Function needed to determine propagation constant.

```
function result = f_TE(z)
% Creates function used to determine propagation constant
% Variable description:
% result - expression used in search for propagation  constant
% z       - actual value of propagation constant
%
% Global variables:
% Global variables are used to transfer values from data
% functions
global n_s          % ref. index substrate
global n_c          % ref. index cladding
global n_layer_int   % ref. index of internal layers
global d_layer_int   % thicknesses of internal layers (microns)
global k_0          % wavenumber
%
zz=z*k_0;
NumLayers = length(d_layer_int);
%
% Creation for substrate and cladding
gamma_sub=sqrt(zz^2-(k_0*n_s)^2);
gamma_clad=sqrt(zz^2-(k_0*n_c)^2);
%
% Creation of kappa for internal layers
kappa=sqrt(k_0^2*n_layer_int.^2-zz.^2);
temp = kappa.*d_layer_int;
%
% Construction of transfer matrix for first layer
cc =  cos(temp);
ss =  sin(temp);
m(1,1) = cc(1);
m(1,2) = -1j*ss(1)/kappa(1);
m(2,1) = -1j*kappa(1)*ss(1);
m(2,2) = cc(1);
%
% Construction of transfer matrices for remaining layers
% and multiplication of matrices
for i=2:NumLayers
    mt(1,1) = cc(i);
```

```
    mt(1,2) = -1j*ss(i)/kappa(i);
    mt(2,1) = -1j*ss(i)*kappa(i);
    mt(2,2) = cc(i);
    m = mt*m;
end
%
result = 1j*(gamma_clad*m(1,1)+gamma_sub*m(2,2))...
      + m(2,1) - gamma_sub*gamma_clad*m(1,2);
```

Listing 12.A.3.3
Function creates one-dimensional mesh.

```
function [N,x_plot,dx,mesh_layer] = mesh_x_CP()
% Generates one-dimensional mesh along x-axis
% Variable description:
% Output
% N              - number of mesh points
% x_plot             - mesh point coordinates used for plotting
% dx             - mesh size
% mesh_layer    - number of mesh points in each layer
%
scaling_dd;
structure;

NumberOfLayers = length(d_layer);  % determine number of layers
delta = d_layer./mesh_layer;
 % separation of points for all layers
%
x(1) = 0.0;                        % coordinate of first mesh point
i_mesh = 1;
for k = 1:NumberOfLayers           % loop over all layers
    for i = 1:mesh_layer(k)        % loop within layer
        x(i_mesh+1) = x(i_mesh) + delta(k);
        i_mesh = i_mesh + 1;
    end
end

N_temp = length(x);
for i=1:N_temp-1
    x_plot(i) = x(i+1);
end

N = length(x_plot);        % number of mesh points

dx = sum(d_layer)/N;       % mesh size
```

```
% Scaling
dx = dx*1d-6;        % Convert mesh size to meters
dx = dx/x_0;         % Scale mesh size
```

Listing 12.A.3.4
Function implements Muller's method.

```
function f_val = muller (f, x0, x1, x2)
% Function implements Muller's method
iter_max = 100;      % max number of steps in Muller method
f_tol   = 1e-6;      % numerical parameters
x_tol = 1e-6;
y0 = f(x0);
y1 = f(x1);
y2 = f(x2);
iter = 0;
while(iter <= iter_max)
    iter = iter + 1;
    a =( (x1 - x2)*(y0 - y2) - (x0 - x2)*(y1 - y2)) / ...
        ( (x0 - x2)*(x1 - x2)*(x0 - x1) );
    %
    b = ( ( x0 - x2 )^2 *( y1 - y2 ) - ( x1 - x2 )^2
            *( y0 - y2 )) / ...
        ( (x0 - x2)*(x1 - x2)*(x0 - x1) );
    %
    c = y2;
    %
    if (a~=0)
        D = sqrt(b*b - 4*a*c);
        q1 = b + D;
        q2 = b - D;
        if (abs(q1) < abs(q2))
            dx = - 2*c/q2;
        else
            dx = - 2*c/q1;
        end
        elseif (b~=0)
            dx = -c/b;
    else
        warning('Muller method failed to find a root')
        break;
    end
    x3 = x2 + dx;
    x0 = x1;
    x1 = x2;
```

```
    x2 = x3;
    y0 = y1;
    y1 = y2;
    y2 = feval(f, x2);
    if (abs(dx) < x_tol && abs (y2) < f_tol)
    break;
    end
end
% Lines below ensure that only proper values are calculated
if (abs(y2) < f_tol)
    f_val = x2;
    return;
else
    f_val = 0;
end
```

Listing 12.A.3.5
Driver function.

```
% File name: slab.m
% Driver function which determines propagation constants and
% electric field profiles (TE mode) for multilayered slab
% structure
clear all;
format long

optical_global;
%
epsilon_M = 1e-6;                  % numerical parameter
TE_mode = [];
n_max = max(n_layer_int);
z1 = n_max;                        % max value of refractive index
n_min = max(n_s,n_c) + 0.001;      % min value of refractive index
dz = 0.005;                        % iteration step
mode_control = 0;
%
while(z1 > n_min)
    z0 = z1 - dz;                  % starting point for Muller method
    z2 = 0.5*(z1 + z0);            % starting point for Muller method
    z_new = muller(@f_TE , z0, z1, z2);
    if (z_new ~= 0)
                                   % veryfying for mode existance
for u=1 : length(TE_mode)
        if(abs(TE_mode(u) - z_new) < epsilon_M)
            mode_control = 1; break; % mode found
```

```
            end
        end
        if (mode_control == 1)
            mode_control = 0;
        else
            TE_mode(length(TE_mode) + 1) = z_new;
        end
    end
    z1 = z0;
end
%
TE_mode = sort(TE_mode, 'descend');
TE_mode'                        % outputs all calculated modes
beta = TE_mode(1);
% selects fundamental mode for plotting field profile
[N,x,d,mesh_layer] = mesh_x_CP();
eps_mesh = epsilon_mesh(mesh_layer);
n_mesh = sqrt(eps_mesh);
TE_mode_field = TE_field(beta,n_mesh,x,k_0);
```

Listing 12.A.3.6
Solves for TE electric field.

```
function TE_mode_field = TE_field(beta,index_mesh,x,k_zero)
% Determines TE optical field for all layers
%
% x - grid created in mesh_x.m
TotalMesh = length(x);  % total number of mesh points
%
zz=beta*k_zero;
%
% Creation of constants at each mesh point
kappa = 0;
for n = 1:(TotalMesh)
    kappa(n)=sqrt((k_zero*index_mesh(n))^2-zz^2);
end
%
% Establish boundary conditions in first layer (substrate).
% Values of the fields U and V are numbered by index not by
% location along x-axis.
% For visualization purposes boundary conditions are
% set at first point.
U(1) = 1.0;
temp = imag(kappa(1));
if(temp<0), kappa(1) = - kappa(1);
```

```
end
% The above ensures that we get a field decaying in the substrate
V(1) = kappa(1);
%
for n=2:(TotalMesh)
   cc=cos( kappa(n)*(x(n)-x(n-1)) );
   ss=sin( kappa(n)*(x(n)-x(n-1)) );
   m(1,1)=cc;
   m(1,2)=-1i/kappa(n)*ss;
   m(2,1)=-1i*kappa(n)*ss;
   m(2,2)=cc;
   %
   U(n)=m(1,1)*U(n-1)+m(1,2)*V(n-1);
   V(n)=m(2,1)*U(n-1)+m(2,2)*V(n-1);
end
%
TE_mode_field = abs(U);                 % Finds Abs(E)
max_value = max(TE_mode_field);
h = plot(x,TE_mode_field/max_value);
% plot normalized value of TE field
% adds text on x-axix and size of x label
xlabel('x (microns)','FontSize',22);
% adds text on y-axix and size of y label
ylabel('TE electric field','FontSize',22);
set(h,'LineWidth',1.5);   % new thickness of plotting lines
set(gca,'FontSize',22);   % new size of tick marks on both axis
pause
close all
```

Bibliography

[1] C. Vassallo. *Optical Waveguide Concepts.* Elsevier, Amsterdam, 1991.

[2] W.P. Huang. Methods for modeling and simulation of guided-wave op-
    toelectronic devices: Part i: Modes and couplings. In J.A. Kong, editor,
    Progress In Electromagnetic Research. PIER 10. Electromagnetic Waves,
    volume 10, pages v–xii. EMW Publishing, Cambridge, Massachusetts,
    1995.

[3] R. Maerz. *Integrated Optics. Design and Modeling.* Artech House, Boston,
    1995.

[4] H. Kogelnik. Theory of optical waveguides. In Theodor Tamir, editor, *Guided-Wave Optoelectronics*, volume 26, pages 7–88. Springer, Berlin, 1998.

[5] A. Yariv. *Quantum Electronics. Third Edition.* Wiley, New York, 1989.

[6] M.S. Wartak. *Computational Photonics. An Introduction with MATLAB.* Cambridge University Press, Cambridge, 2013.

[7] C. Chen, P. Berini, D. Feng, and V.P. Tzolov. Efficient and accurate numerical analysis of multilayer planar optical waveguides. *Proc. SPIE*, 3795:676, 1999.

[8] *Efficient and accurate numerical analysis of multilayer planar optical waveguides*, volume 3795. Proc. SPIE, 1999.

13

Optical Parameters and Processes

Here we summarize optical processes specific to semiconductor lasers and also relevant parameters.

13.1 Optical Parameters

Optical parameters are summarized in Table 13.1. They are: relative dielectric constant ε_r, refractive index $\overline{n}$ related to ε_r by the formula

$$\varepsilon_r = \overline{n}^2 \tag{13.1}$$

and phenomenological parameters which describe optical gain. In the simplest, temperature and wavelength independent linear gain model, there are two of them, a and b. In more complicated models, for example when temperature and/or wavelength dependencies are considered, there will be more parameters needed.

13.1.1 Dielectric function and refractive index

The dielectric function $\varepsilon(\omega)$ describes the response of a material to an electric field $\overrightarrow{E}$. The response depends on the angular frequency ω of the field and frequently results in introducing the low-frequency (static) and high-frequency dielectric constants. In order to include absorption or gain, the complex notation is introduced as

$$\varepsilon(\omega) = \varepsilon'(\omega) + j\varepsilon''(\omega) \tag{13.2}$$

where the real part ε' describes polarization effects and the imaginary part ε'' describes absorption ($\varepsilon'' < 0$) or gain ($\varepsilon'' > 0$).

The complex index of refraction is further written as [1]

$$\overline{n}(\omega) = \overline{n}_r(\omega) + jk_a(\omega) \tag{13.3}$$

DOI: 10.1201/9781003265849-13

TABLE 13.1

Optical parameters.

Symbol	Description
ε_r	*relative dielectric constant*
$\overline{n}$	*refractive index*
a	*linear gain coefficient*
b	*linear gain coefficient*

The real part of refractive index $\overline{n}_r(\omega)$ is determined by the velocity of light in the material as

$$v = \frac{c}{\overline{n}_r} \tag{13.4}$$

where c is the velocity of light in a vacuum. The wavelengths, in a vacuum (λ_0) and the material (λ) are related as

$$\lambda = \frac{\lambda_0}{\overline{n}_r} \tag{13.5}$$

The imaginary part of refractive index is related to the absorption of optical power which is described by the absorption coefficient α_{abs} as

$$k_a = \frac{\lambda \alpha_{abs}}{4\pi} \tag{13.6}$$

The wavelength dependence of the refractive index is modelled very often using Sellmeier formula [2] which is based on the empirical oscillator model. The description applies for photons with energies below the band gap. One of the most popular expressions is the second-order Sellmeier formula [1]

$$\overline{n}^2(\lambda) = s_0 + \frac{s_1 \lambda^2}{\lambda^2 - \lambda_1^2} + \frac{s_2 \lambda^2}{\lambda^2 - \lambda_2^2} \tag{13.7}$$

which uses fitting parameters s_i and λ_i. The values of those parameters for several semiconductors are summarized in Table 4.1 of [1]. Other models are described by [3] and [4].

The effect of the free carrier concentration on the refractive index is often expressed as [5]

$$\overline{n}^2(y) = \overline{n}_0^2(y) - \alpha_R \frac{\overline{n}_0(y)}{k_0} g(n,p) + j \frac{\overline{n}_0(y)}{k_0} [g(n,p) - \alpha_{loss}] \tag{13.8}$$

where $\overline{n}_0$ is the bulk refractive index, α_R is the line-width enhancement factor, k_0 is the wave vector in the vacuum, $g(n,p)$ is the optical gain and α_{loss} are the total losses.

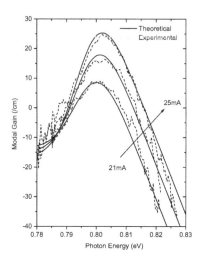

FIGURE 13.1
Fitted theoretical and experimental net modal gain versus photon energy including leakage terms for three values of injected current.

13.1.2   Static permittivity

It is a common practice [6, 7] to use in Poisson equation zero-frequency relative dielectric constant ε_{dc} instead of ε_r.

13.1.3   Optical gain

For efficient and reliable numerical simulations an exact mathematical expression of the material gain is critical. Such expression should also be subject to experimental verification. This problem has been investigated since the early developments of semiconductor lasers [8–10]. The first step is usually the determination of gain spectra and its comparison with experiment. This problem is not discussed here. Typical curves of gain spectra for four quantum well system and comparison with experimental measurements are shown in Fig. 13.1, from [11].

Mathematics involved in determining optical gain of semiconductor quantum-well structures is complex and has been discussed extensively, for example in [3, 12, 13]. As a result analytic approximations of the optical gain which can be used in fast calculations were determined, see [14–16]. From extensive discussions it was determined that the peak material gain for bulk

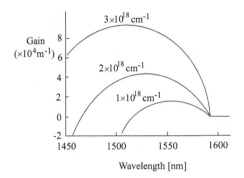

FIGURE 13.2
Schematic optical gain spectra for various carrier densities below and above
the transparency carrier density N_{tr}.

materials vary linearly with carrier density

$$g(n, \lambda) = a(\lambda) \min(n, p) - b(\lambda) \tag{13.9}$$

Here $a(\lambda)$ is commonly called the differential gain.

On the basis of experimental observations, Westbrook [17, 18] extended
the linear gain peak model to allow for wavelength dependence. In its simplest
form it can be written as [17, 19, 20]

$$g(n, \lambda) = a(\lambda_p)n - b(\lambda_p) - b_a (\lambda - \lambda_p)^2 \tag{13.10}$$

where λ_p is the wavelength of the peak gain at transparency (i.e., $g = 0$) and
b_a governs the base width of the gain spectrum. Wavelength peak λ_p can be
also carrier density dependent.

Gain (absorption) in a semiconductor is a function of carrier density n
$[cm^{-3}]$ (see Fig. 13.2, which shows gain spectrum for several values of carrier
concentration) [21]. When n is below *transparency density* n_t, the medium
absorps optical signal. For $n > n_t$, optical gain in the material exceeds loss.
The dependence of the so-called *gain peak* on the carrier density for quantum
well systems is logarithmic, as shown in Fig. 13.3. For modelling purposes, a
linear approximation is often employed.

The gain model employed in this book is given by Eq.(13.9). The coefficients a and b depends on the material and device structure. Their approximate values are provided in later sections.

For systems employing quantum wells the dependence of optical gain
peak versus concentration of carriers is described by the logarithmic function
(see book by Coldren and Corzine [22] for an extensive summary of various
parametrizations). In the simplest form it is

$$g(n) = g_0 \ln \frac{n}{n_{tr}} \tag{13.11}$$

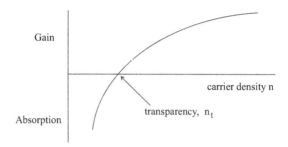

FIGURE 13.3
Logarithmic gain in semiconductor lasers.

where g_0 is the gain coefficient and n_{tr} is the transparency carrier density. In this case the differential gain is $\partial g / \partial n = g_0 / n$.

13.2 Absorption (losses) Coefficients

The expressions used in determining absorption (or gain) are basically the same and within the framework of the first order perturbation theory can be written as [3]

$$\alpha\left(\hbar\omega\right) = \frac{\pi e^2}{\bar{n} c \varepsilon_0 m_0^2 \omega} \frac{2}{V} \sum_{\mathbf{k}} \sum_{i,j(i \neq j)} |\hat{\mathbf{e}} \cdot \mathbf{p}_{ij}|^2 L\left(E_{\mathbf{k}i} - E_{\mathbf{k}j} - \hbar\omega\right) \left[f\left(E_{\mathbf{k}i}\right) - f\left(E_{\mathbf{k}j}\right)\right]$$

(13.12)

where e is the charge of an electron, $\bar{n}$ is the refractive index of the material, c is the speed of light, ε_0 is the permittivity of free space, m_0 is the mass of a free electron, ω is the frequency of the light, V is the volume of the material, i and j are indices that represent one of the three valence bands, $E_{\mathbf{k}i}(E_{\mathbf{k}j})$ and $f\left(E_{\mathbf{k}i}\right)\left[f\left(E_{\mathbf{k}j}\right)\right]$ are the energy and occupation probabilities of initial (final) hole state at the wave vector $\mathbf{k}$, $\hat{\mathbf{e}}$ is the polarization of the light, $\mathbf{p}_{ij}$ is the momentum matrix element between valence band states and L is a broadening function. The sum over i and j represents the three types of transitions, heavy hole to light-hole, heavy-hole to split-off and light-hole to split-off bands.

Equation (13.12) is also used to determine material gain which was summarized in the previous section. However, the two situations are different. Transitions that contribute to gain occur between states in a quite narrow region around the Brillouin zone center ($\mathbf{k} = \mathbf{0}$).

In the model we used phenomenological description of losses and considered three types of mechanisms for internal losses

$$\alpha_{loss} = \alpha_{fc} + \alpha_{ivba} + \alpha_{mirr}$$

(13.13)

TABLE 13.2

Parameters used in determining losses.

Symbol	Description
α_n	*free-carrier absorption coefficient*
α_p	*free-carrier absorption coefficient*
κ_p	*IVBA coefficient*
R	*mirror reflectivity*
L_{cavity}	*cavity length*
α_{abs}	*cavity absorption*

where α_{fc} is responsible for free-carrier absorption, α_{IVBA} is the intervalence band absorption coefficient and α_m describes mirror losses. All losses parameters are summarized in Table 13.2.

13.2.1 Free-carrier absorption

Free-carrier absorption for electrons is a physical process in which an electron absorbs a photon within the conduction band and moves to a higher energy state, Fig. 13.4. As illustrated, both intravalley and intervalley transitions are possible. During the process the electron changes its momentum. In order to obey conservation of momentum, the electron needs to gain an extra momentum by interacting with phonons or impurities. (Photon's momentum is is much smaller and its value cannot provide necessary momentum conservation). The free-carrier absorption is a second-order quantum process which is usually smaller than the first-order band-to-band transition.

Extensive analysis of free-carrier absorption by considering the interactions between electrons and polar optical phonons, deformation potential optical phonons, deformation potential acoustic phonons, piezoelectric acoustic phonons and charged impurities in the intravalley and the intervalley transitions has been reported in [23]. Their analysis is also true for holes if one assumes parabolic band structure for holes.

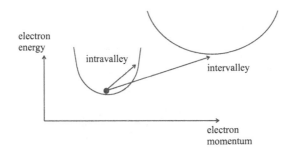

FIGURE 13.4

Schematic diagram of the intravalley and intervalley free-carrier absorption for electrons.

The resulting free-carrier absorption is described by the coefficient α_{fc} which is proportional to concentration of electrons n. Similar mechanism holds for holes, and therefore

$$\alpha_{fc}(n,p) = \alpha_n n + \alpha_p p \qquad (13.14)$$

with two constants α_n and α_p.

13.2.2 Intervalence band absorption

Intervalence band absorption (IVBA) in III-V semiconductors results from radiative transitions in the valence bands, as shown in Fig. 13.5. IVBA is an important process for any photonic device which has regions with large hole concentrations. Large hole concentrations can result due to high p-doping or operation at high carrier injection levels or both. IVBA is the main mechanism of loss in InP based long-wavelength laser diodes [24]. In devices operating at long wavelengths, photons can be absorbed by transitions of holes from one valence band (e.g. heavy-hole band) to another (e.g. spin-off band).

The three possible types of vertical transitions are shown in Fig. 13.5. The light-heavy hole band transition (A) is less important since the transition energies are small. The spin split-off-light-hole band transition (B) is neglected because density of states in a light-hole band is small and a very low hole occupancy of final states for transitions at the bandgap energy [25]. The largest absorption at relevant wavelengths exists for the spin-off-heavy-hole band transitions (C). It is much larger than transitions (B) because larger density of states and hole occupancy in the heavy-hole band.

To model this type of absorption, an appropriate form of Eq.(13.12) has to be evaluated [26]. It is important, however to describe the valence bands

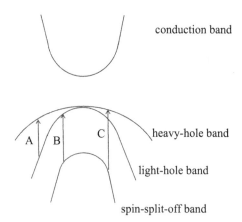

FIGURE 13.5
Intervalence band absorption in III-V semiconductors results from transitions in the valence bands.

accurately for IVBA calculations because transitions take place at large values of wavenumber k. In particular, nonparabolicity and dependence on k direction have to be taken into account.

The intervalence band absorption employed here is given by

$$\alpha_{ivba} = \kappa_p p \tag{13.15}$$

where κ_p is a given parameter.

13.2.3 The mirror loss

The mirror loss is due to the incomplete reflection of the optical wave at the laser facets [27]. It is characterized by the coefficient α_{mirr}. Assuming that R_1 and R_2 are the reflectivities of the two facets, the intensity loss due to reflection at the facets is given by

$$\alpha_{mirr} = \frac{1}{2L_{cavity}} \ln\left(\frac{1}{R_1 R_2}\right) \tag{13.16}$$

where L_{cavity} is the cavity length. If both mirrors have the same reflectivity R, one obtains $\alpha_{mirr} = L_{cavity}^{-1} \ln 1/R$.

13.2.4 Auger processes

These processes were discussed in the earlier chapter. Total Auger recombination rate is expressed as

$$R_{Auger} = (C_n n + C_p p)(np - n_{int}p_{int}) \tag{13.17}$$

where C_n and C_p are the Auger recombination coefficients, respectively, associated with the CHCC and the CHSH Auger process. Auger coefficients C_n and C_p depends on the type of material simulated.

13.3 Spontaneous emission factor

In spontaneous emission the electron falls down in energy from the upper level (E_2) to a lower energy level (E_1) and emits a photon of energy $h\nu = E_2 - E_1$. The photon is emitted in a random direction. The transition is spontaneous provided that the state with energy E_1 is not already occupied by another electron.

The spontaneous emission factor (β_{sp}) is introduced with the photon rate equation [28]

$$\frac{dS}{dt} = v_g(G - \alpha_{tot})S + \beta_{sp}\tilde{R}_{spon} \tag{13.18}$$

with Γ optical confinement factor, R_{st} stimulated emission rate, R_{sp} spontaneous recombination rate and τ_p photon lifetime.

For uniform coupling to all modes, β_{sp} is just the reciprocal of the number of optical modes [22]. The parameter β_{sp} has been discussed extensively in the literature [29], [30, 31]. It is defined as [32]

$$\beta_{sp} = \frac{\text{rate of spontaneous emission into one oscillating mode}}{\text{total rate of spontaneous emission}} \qquad (13.19)$$

For a mode lasing at the wavelength $\lambda = 2\pi c/\omega$, the expression derived by Petermann [32] is

$$\beta_{sp} = \frac{K\Gamma\lambda^4}{4\pi^2 \mu\overline{\mu}\overline{n}_g V \Delta\lambda_{sp}} \qquad (13.20)$$

where μ is the bulk material refractive index, $\overline{\mu}$ group refractive index, $\overline{n}_g$ is the group refractive index, V active volume, K Pettermann enhancement factor and $\Delta\lambda_{sp}$ is the width of the spontaneous-emission spectrum.

In practice β_{sp} is treated as a fitting parameter. Its typical numerical value for index-guided InGaAsP lasers is the range of $10^{-4} - 10^{-5}$ [29]. Its typical experimental value measured with in-plane lasers is [22]

$$\beta_{sp} \approx 1.2 \times 10^{-5} \qquad (13.21)$$

Bibliography

[1] J. Piprek, Cun-Zheng Ning, H. Wunsche, and Siu-Fung Yu. Introduction to the issue on optoelectronic device simulation. *IEEE J. Select. Topics Quantum Electron.*, 9:685–687, 2003.

[2] S.O. Kasap. *Optoelectronics and Photonics: Principles and Practices.* Prentice Hall, Upper Saddle River, New Jersey, 2001.

[3] S.-L. Chuang. *Physics of Photonics Devices.* Wiley, New York, 2009.

[4] S. Adachi. *Physical Properties of III-V Semiconductor Compounds.* Wiley, New York, 1992.

[5] A. Champagne, R. Maciejko, and J.M. Glinski. The performance of double active region InGaAsP lasers. *IEEE J. Quantum Electron.*, 27:2238–2247, 1991.

[6] Z.-M. Li, K.M. Dzurko, A. Delage, and S.P. McAlister. A self-consistent two-dimenional model of quantum-well semiconductor lasers: optimization of a GRIN-SCH SQW laser structure. *IEEE J. Quantum Electron.*, 28:792–803, 1992.

[7] Z-M. Li, M. Dion, S.P. McAlister, R.L. Williams, and G.C. Aers. Incorporating of strain into a two-dimensional model of quantum-well semiconductor lasers. *IEEE J. Quantum Electron.*, 29:346–354, 1993.

[8] G. Lasher and F. Stern. Spontaneous and stimulated recombination radiation in semiconductors. *Phys. Rev.* **A**, 133:553–563, 1964.

[9] C.J. Hwang. Properties of spontaneous and stimulated emission in gaas junction lasers. i. densities of states in the active regions. *Phys. Rev.* **B**, 2:4117–4125, 1970.

[10] F. Stern. Calculated spectral dependence of gain in excited gaas. *J. Appl. Phys.*, 47:5382–5386, 1976.

[11] M.S. Wartak, P. Weetman, T. Alajoki, J. Aikio, V. Heikkinen, N. Pikhin, and P. Rusek. Optical modal gain in multiple quantum well semiconductor lasers. *Canadian Journal of Physics*, 84:53–66, 2006.

[12] J.P. Loehr. *Physics of Strained Quantum Well Lasers*. Kluwer Academic Publishers, Boston, 1998.

[13] P.S. Zory, Jr., editor. *Quantum Well Lasers*. Academic Press, Boston, 1993.

[14] T.-A. Ma, Z.-M. Li, T. Makino, and M.S. Wartak. Approximate optical gain formulas for 1.55 μm strained quaternary quantum well lasers. *IEEE J. Quantum Electron.*, 31:29–34, 1995.

[15] T. Makino. Analytical formulas for the optical gain of quantum wells. *IEEE J. Quantum Electron.*, 32:493 – 501, 1996.

[16] S. Balle. Simple analytical approximations for the gain and refractive index spectra in quantum-well lasers. *Phys. Rev.* **A**, 57:1304–1312, 1998.

[17] L.D. Westbrook. Measurements of dg/dn and dn/dn and their dependence on photon energy in $\lambda = 1.5\mu m$ InGaAsP laser diodes. *IEE Proceedings-part J, Optoelectron.*, 133:135–143, 1986.

[18] L.D. Westbrook. Measurements of dg/dn and dn/dn and their dependence on photon energy in $\lambda = 1.5\mu m$ InGaAsP laser diodes. *IEE Proceedings-part J, Optoelectron.*, 134:122, 1987.

[19] H. Ghafouri-Shiraz and B.S.L. Lo. *Distributed Feedback Laser Diodes*. Wiley, Chichester, 1996.

[20] M.-C. Amann and J. Buus. *Tunable Laser Diodes*. Artech House, Boston, 1998.

[21] M.J. Connelly. *Semiconductor Optical Amplifiers*. Kluwer Academic Publishers, Boston, 2002.

[22] L.A. Coldren and S.W. Corzine. *Diode Lasers and Photonic Integrated Circuits*. Wiley, New York, 1995.

[23] C.-Y. Tsai, C.-Y. Tsai, C-H. Chen, T L. Sung, T-Y. Wu, and F.-P. Shih. Theoretical model for intravalley and intervalley free-carrier absorption in semiconductor lasers: beyond the classical Drude model. *IEEE J. Quantum Electron.*, 34:552–559, 1998.

[24] S. Kakimoto. Intervalence band absorption loss coefficients of the active layer for InGaAs/InGaAsP multiple quantum well laser diodes. *J. Appl. Phys.*, 92:6403–6407, 2002.

[25] G.N. Childs, S. Brand, and R.A. Abram. Intervalence band absorption in semiconductor laser materials. *Semicond. Sci. Technol.*, 1:116–120, 1986.

[26] J. Taylor and V. Tolstikhin. Intervalence band absorption in InP and related materials for optoelectronic device modeling. *J. Appl. Phys.*, 87:1054–1059, 2000.

[27] M. Grundmann. *The Physics of Semiconductors. An Introduction Including Devices and Nanophysics*. Springer, Berlin, 2006.

[28] T. Ohtoshi, K. Yamaguchi, C. Nagaoka, T. Uda, Y. Murayama, and N. Chinone. A two-dimensional device simulator of semiconductor lasers. *Solid State Electron.*, 30:627–638, 1987.

[29] G.P. Agrawal and N.K. Dutta. *Semiconductor Lasers. Second Edition.* Kluwer Academic Publishers, Boston, 2000.

[30] M. Newstein. The spontaneous emission factor for lasers with gain induced waveguiding. *IEEE J. Quantum Electron.*, 21:1270–1276, 1984.

[31] M. Newstein. Further comments. the spontaneous emission factor for lasers with gain induced waveguiding. *IEEE J. Quantum Electron.*, 21:737, 1985.

[32] K. Peterman. *Laser Diode Modulation and Noise*. Kluwer Academic, Dordrecht, 1991.

14

Semiconductor Laser

In this chapter we will describe the practical aspects of numerical implementation of the electrical and optical equations introduced in previous chapters. We concentrate on one-dimensional (1-D) case. Since some of the variables change by many orders of magnitude, it is a common practice to introduce appropriate scaling. After discussing it, other choice of variables will be discussed. Lets start first with the summary of basic electrical equations in 1D along x-axis.

14.1 Summary of Electrical Equations

We first summarize electrical equations in a scaled form.

14.1.1 Poisson equation in equilibrium

In equilibrium Poisson equation involves only potential ψ. After linearization and discretization it can be expressed in a condensed form as

$$a_2(i)\psi_{i+1}^{(n+1)} + b_2(i)\psi_i^{(n+1)} + c_2(i)\psi_{i-1}^{(n+1)} = f^{(n)}(i) \tag{14.1}$$

where

$$a_2(i) = \varepsilon_r(i{-}1), \quad b_2(i) = -\left[2\varepsilon_r(i) + \Delta x^2 \left(e^{-\psi_i^{(n)}} + e^{\psi_i^{(n)}}\right)\right], \quad c_2(i) = \varepsilon_r(i{+}1) \tag{14.2}$$

and

$$f^{(n)}(i) = -\Delta x^2 \left[\left(e^{-\psi_i^{(n)}} - e^{\psi_i^{(n)}} + C_i\right) + \psi_i^{(n)}\left(e^{-\psi_i^{(n)}} + e^{\psi_i^{(n)}}\right)\right] \tag{14.3}$$

14.1.2 Poisson equation in non-equilibrium

In non-equilibrium situation we do not linearize Poisson equation and keep densities of electrons and holes which are updated during calculations. In this case it is expressed in a condensed form as

$$a_2(i)\psi_{i+1}^{(n+1)} + b_2(i)\psi_i^{(n+1)} + c_2(i)\psi_{i-1}^{(n+1)} = f^{(n)}(i) \tag{14.4}$$

DOI: 10.1201/9781003265849-14

where

$$a_2(i) = \varepsilon_r(i-1), \quad b_2(i) = -\left[2\varepsilon_r(i) + \Delta x^2 \left(p_i^{(n)} + n_i^{(n)}\right)\right], \quad c_2(i) = \varepsilon_r(i+1) \tag{14.5}$$

and

$$f^{(n)}(i) = -\Delta x^2 \left[\left(p_i^{(n)} - n_i^{(n)} + C_i\right) + \psi_i^{(n)}\left(p_i^{(n)} + n_i^{(n)}\right)\right] \tag{14.6}$$

14.1.3   Electrons

Equation for electrons

$$a_n(i)n_{i-1} + b_n(i)n_i + c_n(i)n_{i+1} = \Delta x_i R(i) \tag{14.7}$$

where

$$a_n(i) = \mu_n(i - \tfrac{1}{2})B\left(\psi_{i-1} - \psi_i\right) \tag{14.8}$$

$$b_n(i) = -\left[\mu_n(i-\tfrac{1}{2})B\left(\psi_i - \psi_{i-1}\right) + \mu_n(i+\tfrac{1}{2})B\left(\psi_i - \psi_{i+1}\right)\right] \tag{14.9}$$

$$c_n(i) = \mu_n(i+\tfrac{1}{2})B\left(\psi_{i+1} - \psi_i\right) \tag{14.10}$$

14.1.4   Holes

In a similar way one finds for holes (all variables scaled)

$$a_p(i)p_{i-1} + b_p(i)p_i + c_p(i)p_{i+1} = \Delta x_i R(i) \tag{14.11}$$

where

$$a_p(i) = \mu_p(i-\tfrac{1}{2})B\left(\psi_i - \psi_{i-1}\right) \tag{14.12}$$

$$b_p(i) = -\left[\mu_p(i-\tfrac{1}{2})B\left(\psi_{i-1} - \psi_i\right) + \mu_p(i+\tfrac{1}{2})B\left(\psi_{i+1} - \psi_i\right)\right] \tag{14.13}$$

$$c_p(i) = \mu_p(i+\tfrac{1}{2})B\left(\psi_i - \psi_{i+1}\right) \tag{14.14}$$

In the above equations $R(i)$ is the total recombination rate (scaled) at mesh point i.

14.2   Recombination Processes

Here, for completeness we summarized all one-dimensional equations used in defining recombinations. Four types of recombination processes are generally considered in our model and the total recombination rate is

$$R = R_{spon} + R_{stim} + R_{Auger} + R_{SRH} \tag{14.15}$$

R_{spon} describes spontaneous emission, R_{stim} is responsible for stimulated emission, R_{Auger} describes Auger processes, and finally R_{SRH} describes Shockley-Read-Hall recombination. For details, see Chapter 6.

The spontaneous recombination rate is given by

$$R_{spon} = B(np - n_{int}p_{int}) \qquad (14.16)$$

where B is the spontaneous recombination coefficient and n_{int} and p_{int} are the intrinsic densities. Coefficient B has the following form

$$B = B_0 - B_1 \min(n, p) \qquad (14.17)$$

For *AlGaAs* the simpler model with $B = const$ is often used.

The stimulated recombination rate for single mode is given by [1, 2]

$$R_{stim} = g(n, p) v_g S I(x) \qquad (14.18)$$

where S the photon density, $v_g = c/\overline{n}$ the group velocity and $g(n, p)$ the local optical gain. The local optical gain considered here is introduced in a phenomenological way. It is modelled as a linear gain model and is [2, 3] (note that there are slightly different models possible)

$$g = a(N - n_e), \quad N = \min(n, p) \qquad (14.19)$$

The normalized dimensionless optical intensity distribution $I(x)$ for the fundamental optical mode obtained from the solution of the wave equation. It is defined as

$$I(x) = \frac{|E_y(x)|^2}{\int_{-\infty}^{\infty} |E_y(x)|^2 dx} \qquad (14.20)$$

Under steady state conditions, the SRH recombination rate is given by

$$R_{SRH} = \frac{np - n_{int}p_{int}}{\tau_p(n + n_{int}) + \tau_n(p + p_{int})} \qquad (14.21)$$

where τ_p and τ_n are the carrier lifetimes.

Auger recombination was discussed in Chapter 5. The net total Auger recombination rate can be expressed as

$$R_{Auger} = (C_n n + C_p p)(np - n_{int}p_{int}) \qquad (14.22)$$

where C_n and C_p are Auger recombination coefficients for electrons and holes, respectively.

Note: In the present implementation we did not included R_{spon} and R_{Auger}.

TABLE 14.1

Values of recombination and gain coefficients for AlGaAs.

Parameter	Value	Reference
Spontaneous	recombination	
B	$1.52 \times 10^{-10} cm^3/s$	[2]
Gain		
a	$3.32 \times 10^{-16} cm^2$	[2]
n_e	$1.889 \times 10^{18} cm^3$	[2]
v_g	$c/\overline{n}$	
SRH	recombination	
τ_n	5×10^{-5} s	
τ_p	5×10^{-5} s	
Auger	recombination	
C_n	$2.1 \times 10^{-42} m^6/s$	[4]
$C_p = C_n$		[4]

14.2.1   Recombination coefficients

In the present simulator we concentrate on AlGaAs. The list of parameters used to model recombination processes is summarized in a Table 14.1. Scaling of recombination parameters is provided in Table 14.2. We assumed that $S_0 = C_0 = n_{int}$.

14.2.1.1   Scaling

All of the above equations need to be scaled. Scaling factors for recombination terms are summarized in Table 14.3. As before, we have chosen $C_0 = n_{int}$. After scaling they become (index i refers to mesh point, and here we dropped primes which indicate scaling)

$$R_{stim}(i) = v_g a(N_i - n_e)SI(x_i) \qquad (14.23)$$

TABLE 14.2

Definitions of scaling factors for recombination.

Name	Quantity	Definition of scaling
distance	x	$x' = \frac{x}{x_0}$
concentrations	n, p	$n' = \frac{n}{C_0}, p' = \frac{p}{C_0}$
gain parameter	a	$a' = \frac{a}{a_0}$
gain parameter	n_e	$n'_e = \frac{n_e}{C_0}$
group velocity	v_g	$v'_g = \frac{v_g}{v_0}$
photon's density	S	$S' = \frac{S}{S_0}$
electric field	E	$E = \frac{E}{E_0}$

TABLE 14.3
Values of scaling parameters for recombination.

Parameter	Scaling value
a_0	$\frac{1}{v_0 S_0 t_0}$
v_0	c (light velocity)
E_0	$\frac{V_T}{x_0}$
t_0	$\frac{x_0^2}{D_0}$

Here $I(x_i)$ is dimensionless.

$$R_{SRH}(i) = \frac{p_i n_i - 1}{\tau_p\,(n_i + 1) + \tau_n\,(p_i + 1)} \qquad (14.24)$$

14.3 Optical Equations

To describe optical behaviour of laser diode, besides electrical equations one must also solve the wave equation and photon rate equation.

14.3.1 Wave equation

In Chapter 13, we have established formalism which allows us to describe one-dimensional multilayer passive slab waveguides. That part was also tested there. However, in a semiconductor laser, there exist active regions where optical amplification takes place. In those regions refractive index typically depends on the density of electrons and holes. Such modifications are necessary to the previously described formalism and will be described in the present section.

We consider only propagation of TE modes which are described by $(E(x) = E_y(x))$

$$\frac{d^2 E(x)}{dx^2} = \left[\beta^2 - \bar{n}^2(x)k^2\right]\,E(x) \qquad (14.25)$$

Refractive index is

$$\bar{n}^2(x) = \bar{n}_0^2(x) - \alpha_R \frac{\bar{n}_0(x)}{k} g(n,p) + j\frac{\bar{n}_0(x)}{k}\left[g(n,p) - \alpha_{total}\right] \qquad (14.26)$$

where $\bar{n}_0$ is the bulk index of refraction, α_R is the linewidth enhancement factor, $k = 2\pi/\lambda$ is the wavevector in a vacuum, $g(n,p)$ local gain and α_{total} are total losses.

Here, refractive index depends on the electron density n and hole density p. Those densities are determined in the electrical part just described.

TABLE 14.4

Definitions of scaling variables used in wave equation.

Name	Quantity	Definition of scaling
electric field	E	$E' = \frac{E}{E_0}$
wave number	k	$k' = \frac{k}{x_0}$

In the first version of the simulator we neglect the above effect, i.e. assume that refractive index $\bar{n}^2$ in the wave equation is constant.

14.3.1.1 Scaling

All quantities in the above equation must also be scaled. Scaling is done in a similar way as for electrical equations. In Table 14.4 we summarized scaling of all variables appearing in the wave equation. There is some overlap between scaling of electrical variables, since they appear in both cases. After scaling the new equation take the form

$$\frac{d^2 E'(x')}{dx'^2} = x_0^2 \left[\beta^2 - \bar{n}_0^2 k^2 \right] E'(x') \tag{14.27}$$

14.3.2   Photon rate equation

The spontaneous emission factor β_{spon} is introduced with the photon rate equation [5]

$$\frac{dS}{dt} = v_g(\widetilde{G} - \widetilde{\alpha}_{tot})S + \beta_{spon}\widetilde{R}_{spon} \tag{14.28}$$

The tilde ($\widetilde{}$) symbol refers to appropriate quantities which are interpreted with the profile of the intensity of electric field. The quantities are defined as follows:

- the total gain is

$$\widetilde{G} = \int_{-\infty}^{\infty} dx \; g(n,p) \; I(x) \tag{14.29}$$

  where $g(n,p)$ is given by Eq. (14.19).

- the total losses are

$$\widetilde{\alpha}_{tot} = \alpha_{mirr} + \alpha_{abs} + \int_{-\infty}^{\infty} dx \; [\alpha_{fc}(n,p) + \alpha_{ivba}(p)] \; I(x) \tag{14.30}$$

- the spontaneous emission rate is

$$\widetilde{R}_{spon} = \int_{-\infty}^{\infty} dx R_{spon}(x) I(x) \tag{14.31}$$

TABLE 14.5

Parameters for losses.

Parameter	Value	Reference
Free-carrier	absorption	
α_n	$3 \times 10^{-18} cm^2$	[4]
α_p	$7 \times 10^{-18} cm^2$	[4]

In the above expressions there is an electric field which need to be determined. It is done through solution of the wave equation with refractive index dependent on electron and hole concentrations.

The free-carrier absorption is described by the coefficient α_{fc} which is proportional to concentration of electrons n. Similar mechanism holds for holes, and therefore

$$\alpha_{fc}(n,p) = \alpha_n n + \alpha_p p \qquad (14.32)$$

with two constants α_n and α_p.

The intervalence band absorption employed here is given by

$$\alpha_{ivba} = \kappa_p p \qquad (14.33)$$

It is not included in the first version of the simulator.

The mirror loss, α_m, assuming the same mirror reflectivities at both ends is given by

$$\alpha_{mirr} = \frac{1}{L_{cavity}} \ln\left(\frac{1}{R}\right) \qquad (14.34)$$

where L_{cavity} is the cavity length and R is the facets reflectivity. We used $R = 0.31$ as a default value. This value can be changed by the user.

Values of parameters of losses are summarized in a Table 14.5.

14.3.2.1 Scaling

All the variables appearing in Eq. (14.28) must also be scaled. Definitions of scaling factors were summarized earlier. In the following we are not including IVBA mechanism. Photon density is scaled as $S' = S/S_0$ with the choice $S_0 = 1 \times 10^{20} m^{-3}$ [6].

The scaled rate equation is

$$\frac{dS'}{dt'} = v_g'(\widetilde{G}' - \widetilde{\alpha}'_{tot})S' \qquad (14.35)$$

The factors are

$$\widetilde{G}' = \int_{-\infty}^{\infty} dx \; g'(n',p') \, I(x) \qquad (14.36)$$

$$\widetilde{\alpha}'_{tot} = \alpha_{mirr} + \alpha_{abs} + \int_{-\infty}^{\infty} dx \; [\alpha_{fc}(n,p) + \alpha_{ivba}(p)] \, I(x) \qquad (14.37)$$

14.3.3 Output power

The expression for the output power per facet which is given by [7]

$$P = \frac{1}{2} (\hbar\omega) S v_g \alpha_{mirr} V_v ol \tag{14.38}$$

where $V_v ol$ is the laser cavity volume and ω the angular frequency of the emitted light.

Here we follow Coldren and Corzine [8] and use similar expression

$$P =\sim \eta_0 \frac{h\nu}{\tau_{ph}} S V_{ph} \tag{14.39}$$

where V_{ph} is the volume occupied by photons.

Photon density is [8]

$$S = \frac{\Gamma R_{sp}}{1/\tau_{ph} - \Gamma v_g g} \tag{14.40}$$

14.3.4 Practical photon rate equation

For practical implementation we used another form of photon rate equation

$$\frac{dS}{dt} = v_g (G - \tau_{ph}^{-1}) S \tag{14.41}$$

where τ_{ph} being photon life-time, $\tau_{ph} = 2.77ps$ [8].

14.4 Remaining Material Parameters

Below is a list of remaining parameters used for $Al_x Ga_{1-x} As$.

14.4.1 Static permittivity

Static permittivity for $Al_x Ga_{1-x} As$ is

$$\varepsilon_{dc} = (13.1 - 3x) \varepsilon_0 \tag{14.42}$$

14.4.2 Carrier mobilities

For mobilities of electrons and holes for $Al_x Ga_{1-x} As$ we use the same models as before, from Shur [9]

$$\mu_n = \left(80000 - 22000x + 10000x^2\right) \frac{cm^2}{V \ s} \tag{14.43}$$

$$\mu_p = \left(370 - 970x + 740x^2\right) \frac{cm^2}{V \ s} \tag{14.44}$$

14.5 Description of the Program

As was indicated earlier, we only present implementation for nondegenerate situation (for Boltzmann statistics). We also use "natural" variables n, p, ψ. There are possibilities of implementing simulator using different set of variables as discussed earlier. Also, for degenerate case where one must apply Fermi-Dirac statistics it is possible to use Boltzmann-like formulation.

The solutions of separate electrical and optical problems were described in the previous chapters. For electrical problem, we analyzed heterostructure system and considered electrical current over heterostructure regions without propagation of light.

For optical problem we only analyzed mode propagation for passive waveguide, i.e. where refractive index does not depend on concentrations of carriers and where one does not have an amplification.

Full case, i.e. simulator with both optical propagation and electrical transport is described next. Schematic flow of the simulator is illustrated in Fig. 14.1.

We assumed that all quantities are scaled and to avoid unnecessary indices we will drop superscript. We will also restrict further developments to the d.c. case and therefore assume that all time dependencies vanish, i.e. $\frac{\partial n}{\partial t} = \frac{\partial p}{\partial t} = \frac{\partial S}{\partial t} = 0$. We also set $G_{gen} = 0$.

The system of equations we attempt to solve is a nonlinear one. Good trial values (initial guesses) are required. The quality of initial guess plays significant role in the numerical procedure and strongly influence amount of time required to find solution. Several methods have been invented, see [10].

14.5.1 Electrical part

This part is based on files developed to model heterostructure, including p-i-n junction. It was extensively described earlier.

14.5.2 Optical part

Optical part was discussed in the earlier chapter. The program was also tested there.

In practical simulations determination of mode propagation constant is critical and one would always like to have an independent verification of the routines used to find propagating modes. In the routines used here we only look for fundamental mode for each waveguide structure. If needed one can always generalize those routines and to find other modes.

Results of optical tests were discussed in a separate chapter. Here we only give detailed description of the routines used. This part of the program works separately without all other files. Their main purpose is to test optical routines

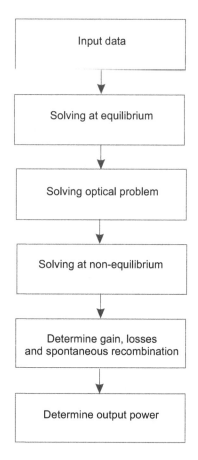

FIGURE 14.1
Schematic flow of the simulator.

by determining propagation constant and field distribution for fundamental
mode. Several structures described in literature were tested by us and compared with the published date. Excellent agreement has been found.

There is some usage of global variables. They are used to transfer data into
corresponding numerical routines which otherwise will be difficult to achieve.
To distinguish those variables, I added word global at the end of each name.
It indicates that the corresponding variable is used in several m-files.

14.5.3   Full simulator

Main program is a collection of files able to simulate semiconductor laser
based on phenomenological parameters. It contains m-files divided into several
groups according to performed functionalities.

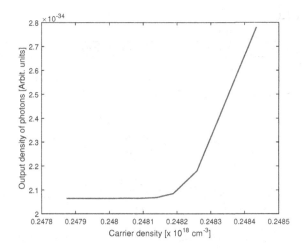

FIGURE 14.2
Output density of photons vs carrier density.

In Table 14.6 we show all files along with brief description. The m-files are classified according to tasks performed.

14.6   Results of Simulations

We first determined values of applied voltage V and current I at threshold and analyzed how photon density S affects current at threshold I_{th}.

Next we used an approach by Coldren and Corzine [8] to determine output power. We assumed an arbitrarily value of photon density $S = 1 \times 10^9 cm^{-3}$.

This is consistent with the values reported in literature. Colderen et al [8] for in-plane device with thickness of quantum well of $80 Angstroms$ reported $S = 2.43 \times 10^{14} cm^{-3}$, whereas Hakki [6] for an active layer with thickness of $0.2 \mu m$ reported $S_{th} = 2 \times 10^{12} cm^{-3}$ at threshold. Some results are shown in Fig. 14.2.

Appendix: Newton Method

For a future extensions we summarize here Newton method which can be applied to analyze the present case.

TABLE 14.6

List of m-files for full simulator.

Listing	Function name	Description
Data files		
14.A.1.1	*param_3_layer_laser.m*	parameters used pin structure
14.A.1.2	*struct_3_layer_pin.m*	pin structure
General		
14.A.2.1	*ber.m*	*numerical implementation of Bernoulli function*
14.A.2.2	*mesh_x_CP.m*	creates mesh
14.A.2.3	*optical_global.m*	transfer of global variables and parameters
14.A.2.4	*physical_const.m*	*contains general physical constants*
14.A.2.5	*scaling_dd.m*	keeps all scaling factors
Models		
14.A.3.1	*gain_linear_AlGaAs.m*	*data for linear gain model at single mesh point*
14.A.3.2	*gain_mesh.m*	*gain for all mesh points*
14.A.3.3	*mobility_n.m*	*mobility of electrons*
14.A.3.4	*mobility_p.m*	*mobility of holes*
14.A.3.5	*recombination_spon.m*	*spontaneous emission at a single mesh point*
14.A.3.6	*recombination_SRH.m*	*SHR recombination*
14.A.3.7	*recombination_stim.m*	*stimulated emission at a single mesh point*
Optical field		
14.A.4.1	*epsilon_mesh.m*	assigns ε_r for all mesh points
14.A.4.2	*f_TE.m*	use to determine electric field
14.A.4.3	*muller.m*	implements Muller method
14.A.4.4	*rate_eq.m*	determines gain and losses
14.A.4.5	*slab_laser.m*	controls optical calculations
14.A.4.6	*TE_field.m*	solves for TE electric field
Semiconductors		
14.A.5.1	*boundary_psi.m*	b.c. for potential
14.A.5.2	*currents.m*	Determines currents
14.A.5.4	*doping_mesh.m*	Assigns doping for all mesh points
14.A.5.3	*electron_density_2.m*	Determines density of electrons
14.A.5.5	*equilibrium_april.m*	Determines quantities in equilibrium
14.A.5.6	*hole_density_2.m*	Determines density of holes
14.A.5.7	*initial.m*	Determines initial potential
14.A.5.8	*nonequilibrium_april.m*	Determines quantities in nonequilibrium
14.A.5.9	*pin_hetero_CP.m*	Driver
14.A.5.10	*potential_eq_feb.m*	Determines potential in equilibrium
14.A.5.11	*potential_noneq_2.m*	Determines potential in non-equilibrium

Our system of equations comprise a highly nonlinear problem. Within this problem, there is a need to determine photon density for a given conditions. Therefore photon rate equation is used to control execution of the program.

For that purpose a Newton method is used. In short, the method is formulated as finding x^* from

$$f(x^*) = 0$$

where $f(x)$ is some nonlinear function of the variable x. The standard one-dimensional algorithm is based on the following equation (here index n refers to iteration number)

$$x^{(n+1)} = x^{(n)} - \frac{f(x^{(n)})}{f'(x^{(n)})} \tag{14.45}$$

which is used to determine next approximation $x^{(n+1)}$ from knowledge of the previous on $x^{(n)}$. For our problem at hand, we use rate equation which is

$$\frac{dS}{dt} = F(S) \tag{14.46}$$

where S is photon density (we assume single-mode operation) and

$$F(S) = v_g \left(G - \overline{\alpha} \right) S \tag{14.47}$$

In the steady-state, i.e. $\frac{dS}{dt} = 0$, one has $F(S^*) = 0$ where S^* is the photon density corresponding to steady-state. Applying Eq.(..), the next iteration value of S is obtained as

$$S^{(n+1)} = S^{(n)} - \frac{F(S^{(n)})}{\frac{F(S^{(n)}) - F(S^{(n-1)})}{S^{(n)} - S^{(n-1)}}} \tag{14.48}$$

where we have made a standard numerical approximation to the derivative of the function $F(S)$.

Appendix 14A: MATLAB Listings

14.6.1   Data files

Listing 14.A.1.1
Contains parameters for p-i-n structure.

```
% File name: param_3_layer_laser.m
% Structure based on AlGaAs
%----------------------------------------------------------------
scaling_dd;
%----------------------------------------------------------------
% Aluminum composition in different layers
```

```
% Al_x Ga_1-x As
%-------------------------------------------------------------
x_comp = 0.25;
%
%-------------------------------------------------------------
% Values of dielectric constant in each layer
% epsilon = (13.1 - 3x)*eps_zero
%-------------------------------------------------------------
eps_layer(3)   = 13.1 - 3*x_comp;
eps_layer(2)   = 13.1;
eps_layer(1)   = 13.1 - 3*x_comp;
%-------------------------------------------------------------
n_layer = sqrt(eps_layer);
%-------------------------------------------------------------
% Assigns value of mobility for each layer
% Mobility does not depend on electric field
% Simple model of mobilities of electrons for Al_x Ga_(1-x) As
% Based on Shur, p.626
mu_n_layer(3) = 80000-22000*x_comp +10000*x_comp*x_comp;   % [cm^2/(V s]
mu_n_layer(2) = 80000;  % [cm^2/(V s]
mu_n_layer(1) = 80000-22000*x_comp +10000*x_comp*x_comp;   % [cm^2/(V s]
mu_n_layer = mu_n_layer*1d-4;      % convert to [m^2/V s]
mu_n_layer = mu_n_layer/mu_0;              % scaling

mu_p_layer(3) = 370-970*x_comp +740*x_comp*x_comp;  % [cm^2/(V s]
mu_p_layer(2) = 370;  % [cm^2/(V s]
mu_p_layer(1) = 370-970*x_comp +740*x_comp*x_comp;  % [cm^2/(V s]
mu_p_layer = mu_p_layer*1d-4;      % convert to [m^2/V s]
mu_p_layer = mu_p_layer/mu_0;              % scaling
%-------------------------------------------------------------
E_0 = 5;                              % reference energy [eV]
chi_e_GaAs = 4.07;                    % Electron affinity for GaAs [eV]
E_g_GaAs = 1.42;                      % Bandgap of GaAs [eV]
chi_e_AlGaAs = 4.07-1.1*x_comp;       % Electron affinity for AlGaAs [eV]
E_g_AlGaAs = 1.424+1.427*x_comp;      % Bandgap of AlGaAs [eV]
band_offset = 0.67;
%--------------------------------
chi_layer(3) = chi_e_GaAs;
chi_layer(2) = chi_e_AlGaAs;
chi_layer(1) = chi_e_GaAs;
E_g_layer(3) = E_g_GaAs ;
E_g_layer(2) = E_g_AlGaAs;
E_g_layer(1) = E_g_GaAs ;
%-------------------------------------------------------------
% Doping values in engineering units
N_A = -1d17;              % acceptor doping [cm^-3]
N_D = 1d18;               % donor doping [cm^-3]
N_neut = 1d15;            % undoped region [cm^-3]
% Convert to SI system
```

```
N_A = N_A*1d6;              % [m^3]
N_D = N_D*1d6;              % [m^3]
N_neut = N_neut*1d6;       % [m^-3]
%----------------------------------------------------------------
dop_layer(3) = N_D;         % in m^-3
dop_layer(2) = N_neut;       % in m^-3
dop_layer(1) = N_A;         % in m^-3
%------------- Scaling of data ------------------------
dop_layer = dop_layer/C_0;
%----------------------------------------------------------------
% Definition of active regions (with gain)
gain_layer(3) = 0;          % no gain
gain_layer(2) = 1;          % active region
gain_layer(1) = 0;          % no gain
```

Listing 14.A.1.2
Defines p-i-n structure.

```
% File name: struct_3_layer_pin.m
% Input needed to construct mesh
% Contains data.
% Global variables to be transferred to function f_TE.m

% Thicknesses of each layer; values in microns
d_layer(3)   = 3.0;  % microns
d_layer(2)   = 0.5;  % microns
d_layer(1)   = 3.0;  % microns

d_layer_int = d_layer(2);    % thickness of internal layer

mesh_layer = [5 3 5];  % number of mesh points in each layer
%mesh_layer = [30 30 30];

lambda     = 1.3;  % wavelength in microns
k_0        = 2*pi/lambda;
```

14.6.2   General

Listing 14.A.2.1
Numerical implementation of Bernoulli function

```
% Simplw implementation of Bernouli function

 function b = ber(x)
```

```
  if abs(x) >= 1.0d-4
    b = x/(exp(x)-1.0);
  else
    b = 1.0 - 0.5*x*(1.0 + x/6.0)*(1.0-x*x/60.);
  end

end
```

Listing 14.A.2.2
Creates one-dimensional mesh.

```
function [N,x_plot,dx,mesh_layer] = mesh_x_CP()
% Generates one-dimensional mesh along x-axis
% Variable description:
% Output
% N              - number of mesh points
% x_plot              - mesh point coordinates used for plotting
% dx             - mesh size
% mesh_layer     - number of mesh points in each layer
%
scaling_dd;
struct_3_layer_pin;

NumberOfLayers = length(d_layer);  % determine number of layers
delta = d_layer./mesh_layer;
% separation of points for all layers
%
x(1) = 0.0;                        % coordinate of first mesh point
i_mesh = 1;
for k = 1:NumberOfLayers           % loop over all layers
    for i = 1:mesh_layer(k)        % loop within layer
        x(i_mesh+1) = x(i_mesh) + delta(k);
        i_mesh = i_mesh + 1;
    end
end

N_temp = length(x);
for i=1:N_temp-1
    x_plot(i) = x(i+1);
end

N = length(x_plot);     % number of mesh points

dx = sum(d_layer)/N;    % mesh size
% Scaling
```

```
dx = dx*1d-6;        % Convert mesh size to meters
dx = dx/x_0;         % Scale mesh size
```

Listing 14.A.2.3
Controls transfer of global variables and parameters used by optical part.

```
% File name: optical_global.m
% Contains variables needed in optical calculations
% Global variables to be transferred to function f_TE.m
% Contains data for 1D waveguide.

global n_c              % ref. index cladding
global n_layer_int       % ref. index of internal layers
global n_s              % ref. index substrate
global d_c              % thickness of cladding (microns)
global d_layer_int
% thicknesses of internal layers (microns)
global d_s              % thickness of substrate (microns)
global k_0              % wavenumber
%
struct_3_layer_pin;
% Construct structure required in optical calculations
N_layers = length(d_layer); % determines number of layers
d_layer_int = [d_layer(2)]; % only internal layers
d_c = d_layer(N_layers);
d_s = d_layer(1);
%
% Establish values of refractive indices
param_3_layer_AlGaAs_laser;
n_layer_int = n_layer(2);
n_c = n_layer(N_layers);
n_s = n_layer(1);
```

Listing 14.A.2.4
Contains general physical constants.

```
% File name: physical_const.m
% Definitions of basic physical constants

eps_zero  = 8.85d-12;  % Permittivity of free space [F/m]
q       = 1.60d-19;  % elementary charge [C]
k_B     = 1.38d-23;  % Boltzmann constant [J/K]
T       = 300;       % Temperature [K]
c_light = 3d8;       % velocity of light [m/s]
n_int   = 1.5d16;    % Intrinsic carrier concentration [m^-3]
```

```
ref_index = 3.4;          % Typical value of refractive index
```

Listing 14.A.2.5
Keeps all scaling factors.

```
% File name: scaling_dd.m
% Purpose: keeps general and electrical scaling factors for dd
% model
% Variable description:
% Basic scaling:
% V_T      - thermal voltage
% x_zero   - length scaling
% C_zero   - concentration scaling
% D_zero   - diffusion scaling
%
% Derived scaling:
% lambda   - Poisson equation scaling
% t_zero   - time scaling
% R_zero   - recombination scaling
% J_zero   - current scaling
% mu_zero  - mobility scaling
% v_0      - scaling of group velocity
% g_0      - gain scaling

physical_const;           % needed for V_T, n_int

V_T = k_B*T/q;            % Thermal voltage[eV]
L_Dint = sqrt(eps_zero*V_T/(q*n_int));

x_0 = L_Dint;            % unit [meters]
C_0 = n_int;             % unit [m^-3]
S_0 = 1d20;      % Scaling of photon density [m^-3]
D_0 = 1d3;          % [cm^2/s]
v_0 = c_light;       % scaling of group velocity
t_0 = (x_0^2)/D_0;
mu_0 = D_0/V_T;
g_0 = 1/(t_0*v_0);       % scaling for gain
a_0 = 1/(v_0*S_0*t_0);
```

14.6.3 Models

Listing 14.A.3.1
Defines data for linear gain model.

```
function g = gain_linear_AlGaAs(p,n)
% File name: gain_linear_AlGaAs.m
% Purpose: Determines optical gain based on phenomenological
% model.
%           Linear gain model is used.
% Gain is determined for all mesh points
% Input n,p are NOT scaled
% Output gain is scaled
%
% Values of parameters for GaAs from
% G-L. Tan et al, JQE, 29, 822 (1993)

scaling_dd;

a   = 3.32d-16;   % unit cm^2
n_e = 1.889d18;   % unit cm^-3
%
% Quantities are scaled as: a' = a*a_0, n_e' = n_e/C_0,
% N' = N/C_0
%
a = a/a_0;
n_e = n_e/C_0;

g = a*(min(p,n) - n_e);   % linear gain model

% if g < 0
%    g = 0;
end
```

Listing 14.A.3.2
Defines gain for all mesh points.

```
function gg = gain_mesh(mesh_layer,n,p)
%-------------------------------------------------------------------
% File name: gain_mesh.m
% Purpose:   Assigns the value of gain for all mesh points.
%            That value is nonzero for the active region.
%            For other layers it is zero

% mesh_layer - number of mesh points in each layer
% gg_mesh  - determines if given mesh point belongs to an
% active layer

param_3_layer_AlGaAs_laser;
```

```
NumberOfLayers = length(gain_layer);
%
i_mesh = 1;
for k = 1:NumberOfLayers                % loop over all layers
    for i = 1:mesh_layer(k)
    % loop within layer over mesh points
        gg_mesh(i_mesh) = gain_layer(k);
        % gain_layer - defines active region
        i_mesh = i_mesh + 1;
    end
end

dd = gain_linear_AlGaAs(p,n)';  % determine gain for all
% mesh points

gg = gg_mesh.*dd;  % gain for mesh points within active region
```

Listing 14.A.3.3
Defines mobility of electrons.

```
function mu_mesh = mobility_n_AlGaAs_mesh(mesh_layer)
% File name: mobility_n_AlGaAs_mesh.m
% Purpose:   Assigns the value of mobility for AlGaAs for all
% mesh points.
%            That value stays the same within each layer and
%            changes from layer to layer.
% mesh_layer)   - number of mesh points in layers
% mu_mesh       - keeps mobility at each mesh point
%
param_3_layer_AlGaAs_laser;

NumberOfLayers = length(eps_layer);
%
i_mesh = 1;
for k = 1:NumberOfLayers                % loop over all layers
    for i = 1:mesh_layer(k)             % loop within layer
        mu_mesh(i_mesh) = mu_n_layer(k);
        i_mesh = i_mesh + 1;
    end
end
```

Listing 14.A.3.4
Defines mobility of holes.

```
function mu_mesh = mobility_p_AlGaAs_mesh(mesh_layer)
```

```
% File name: mobility_n_AlGaAs_mesh.m
% Purpose:   Assigns the value of mobility for AlGaAs for all
% mesh points.
%            That value stays the same within each layer and
%            changes from layer to layer.
% mesh_layer     - mobility of holes within each layer
% mu_mesh        - keeps mobility at each mesh point
%
param_3_layer_AlGaAs_laser;

NumberOfLayers = length(eps_layer);
%
i_mesh = 1;
for k = 1:NumberOfLayers              % loop over all layers
    for i = 1:mesh_layer(k)           % loop within layer
        mu_mesh(i_mesh) = mu_p_layer(k);
        i_mesh = i_mesh + 1;
    end
end
```

Listing 14.A.3.5
Defines spontaneous recombination

```
function r_spon = recombination_spon(n,p)
% Determines spontaneous recombination for all mesh points
%at mesh point 'i'
% n,p,S  - all scaled

physical_const;
scaling_dd;

B = 1.52d-10;   % unit cm^3/s
B = B*t_0/C_0;
r_spon = B*(n.*p - 1);

end
```

Listing 14.A.3.6
Defines SRH recombination.

```
function rec_SRH = recombination_SRH(i,dop,n,p)
% Determines recombination at mesh point 'i'

scaling_dd
```

```
tau_n0 = 1e-5;  % SRH recombination time for electrons [seconds]
tau_p0 = 1e-5;  % SRH recombination time for holes [seconds]
N_ref_n = 5e16;
N_ref_p = 5e16;

NN = 2*abs(dop(i));
tau_n = tau_n0/(1 + NN/N_ref_n);      % electron lifetime
tau_p = tau_p0/(1 + NN/N_ref_p);      % hole lifetime

rnum  = n(i)*p(i) - 1;
denom = tau_n*(p(i)+1)+tau_p*(n(i)+1);
denom = denom/t_0;         % scaling
rec_SRH   = (rnum/denom);
end
```

Listing 14.A.3.7
Defines stimulated recombination.

```
function r_stim = recombination_stim(i,n,p,S,I_int)
% Determines stimulated recombination at mesh point 'i'
% n,p,S  - all scaled

physical_const;
scaling_dd;

v_g = c_light/ref_index;    % definition of group velocity
v_g = v_g/v_0;              % scaling of group velocity

g = gain_linear_AlGaAs(p(i),n(i));
r_stim = g*v_g*S*I_int(i);
% 'in stim'
% g
% r_stim
% pause

end
```

14.6.4   Optical field

Listing 14.A.4.1
Assigns ε_r for all mesh points.

```
function eps_mesh = epsilon_mesh(mesh_layer)
%----------------------------------------------------------------
```

```
% File name: epsilon_mesh.m
% Purpose:    Assigns the value of dielectric constant for all
% mesh points.
%          That value stays the same within each layer and changes
%          from layer to layer.
% eps_layer - dielectric constants in each layer
% eps_mesh  - keeps epsilon_dc for each mesh point

param_3_layer_AlGaAs_laser;  % provides data for eps_layer

NumberOfLayers = length(eps_layer);
%
i_mesh = 1;
for k = 1:NumberOfLayers              % loop over all layers
    for i = 1:mesh_layer(k)           % loop within layer
        eps_mesh(i_mesh) = eps_layer(k);
        i_mesh = i_mesh + 1;
    end
end
```

Listing 14.A.4.2
Determines electric field.

```
function result = f_TE(z)
% Creates function used to determine propagation constant
% Variable description:
% result - expression used in search for propagation  constant
% z      - actual value of propagation constant
%
% Global variables:
% Global variables are used to transfer values from data
% functions
global n_s          % ref. index substrate
global n_c          % ref. index cladding
global n_layer_int    % ref. index of internal layers
global d_layer_int    % thicknesses of internal layers
% (microns)
global k_0          % wavenumber
%
zz=z*k_0;
NumLayers = length(d_layer_int);
%
% Creation for substrate and cladding
gamma_sub=sqrt(zz^2-(k_0*n_s)^2);
gamma_clad=sqrt(zz^2-(k_0*n_c)^2);
```

```
%
% Creation of kappa for internal layers
kappa=sqrt(k_0^2*n_layer_int.^2-zz.^2);
temp = kappa.*d_layer_int;
%
% Construction of transfer matrix for first layer
cc =  cos(temp);
ss =  sin(temp);
m(1,1) = cc(1);
m(1,2) = -1j*ss(1)/kappa(1);
m(2,1) = -1j*kappa(1)*ss(1);
m(2,2) = cc(1);
%
% Construction of transfer matrices for remaining layers
% and multiplication of matrices
for i=2:NumLayers
    mt(1,1) = cc(i);
    mt(1,2) = -1j*ss(i)/kappa(i);
    mt(2,1) = -1j*ss(i)*kappa(i);
    mt(2,2) = cc(i);
    m = mt*m;
end
%
result = 1j*(gamma_clad*m(1,1)+gamma_sub*m(2,2))...
        + m(2,1) - gamma_sub*gamma_clad*m(1,2);
```

Listing 14.A.4.3
Implements Muller method

```
function f_val = muller (f, x0, x1, x2)
% Function implements Muller's method
iter_max = 100;       % max number of steps in Muller method
f_tol    = 1e-6;      % numerical parameters
x_tol = 1e-6;
y0 = f(x0);
y1 = f(x1);
y2 = f(x2);
iter = 0;
while(iter <= iter_max)
    iter = iter + 1;
    a =( (x1 - x2)*(y0 - y2) - (x0 - x2)*(y1 - y2)) / ...
        ( (x0 - x2)*(x1 - x2)*(x0 - x1) );
    %
    b = ( ( x0 - x2 )^2 *( y1 - y2 ) - ( x1 - x2 )^2
    *( y0 - y2 ))
```

```
       /  ...
     (  (x0 - x2)*(x1 - x2)*(x0 - x1) );
  %
  c = y2;
  %
  if (a~=0)
       D = sqrt(b*b - 4*a*c);
       q1 = b + D;
       q2 = b - D;
       if (abs(q1) < abs(q2))
            dx = - 2*c/q2;
       else
            dx = - 2*c/q1;
       end
       elseif (b~=0)
            dx = -c/b;
  else
       warning('Muller method failed to find a root')
       break;
  end
  x3 = x2 + dx;
  x0 = x1;
  x1 = x2;
  x2 = x3;
  y0 = y1;
  y1 = y2;
  y2 = feval(f, x2);
  if (abs(dx) < x_tol && abs (y2) < f_tol)
  break;
  end
end
% Lines below ensure that only proper values are calculated
if (abs(y2) < f_tol)
    f_val = x2;
    return;
else
    f_val = 0;
end
```

Listing 14.A.4.4
Determines gain and losses

```
function [F_out,g1,a1,tot_den_n,tot_den_p,R_spon_bar]
          = rate_eq(x,I_int,n,p,S)
% It determines function F(S)
```

```
% Function controls determination of gain and losses for all
% mesh points

scaling_dd;
struct_3_layer_pin;     % needed for mesh_layer
physical_const;         % needed for c_light and ref_index

% Extra calculations
tot_den_n = trapz(x,n);

r_spon = recombination_spon(n,p);

cc = r_spon'.*I_int;
R_spon_bar = trapz(x,cc);

tot_den_p = trapz(x,p);

g_mesh = gain_mesh(mesh_layer,p,n);
% gain at all mesh points !!!!!!

yy = g_mesh.*I_int;     % Calculate modal gain

G_bar = trapz(x,yy);    %/(v_0*t_0);  % integration of modal gain

v_g = c_light/ref_index;
v_g = v_g/v_0;          % scaling of group velocity

% Determination of function F(S), using photon lifetime
tau_ph = 2.77*1d-9/t_0;
F_out_ph = v_0*(G_bar - tau_ph);
F_out = F_out_ph*S;     % Here S is already scaled
g1 = G_bar;
a1 = tau_ph;

end
```

Listing 14.A.4.5
Controls oprical calculations.

```
function [I_int] = slab_laser(x,mesh_layer)

% File name: slab_laser.m
% Function which determines propagation constants and
% electric field profiles (TE mode) for multilayered slab
% structure
```

```
%clear all;
format long

optical_global;
%
epsilon_M = 1e-6;                  % numerical parameter
TE_mode = [];
n_max = max(n_layer_int);
z1 = n_max;                        % max value of refractive index
n_min = max(n_s,n_c) + 0.001;      % min value of refractive index
dz = 0.005;                        % iteration step
mode_control = 0;
%
while(z1 > n_min)
    z0 = z1 - dz;                  % starting point for Muller method
    z2 = 0.5*(z1 + z0);            % starting point for Muller method
    z_new = muller(@f_TE , z0, z1, z2);
    if (z_new ~= 0)
                            % veryfying for mode existance
for u=1 : length(TE_mode)
        if(abs(TE_mode(u) - z_new) < epsilon_M)
            mode_control = 1; break; % mode found
        end
    end
    if (mode_control == 1)
        mode_control = 0;
    else
        TE_mode(length(TE_mode) + 1) = z_new;
    end
    end
    z1 = z0;
end
%
TE_mode = sort(TE_mode, 'descend');
%TE_mode'                          % outputs all calculated modes
beta = TE_mode(1);
% selects fundamental mode for plotting field profile
%[N,x,d,mesh_layer] = mesh_x_CP();
eps_mesh = epsilon_mesh(mesh_layer);
n_mesh = sqrt(eps_mesh);
TE_mode_field = TE_field(beta,n_mesh,x,k_0);
% field profile for whole waveguide

% Determines normalized intensity for all mesh points
field_abs = abs(TE_mode_field);
```

```
field_abs2 = field_abs.^2;
I_zero = trapz(x,field_abs2);
I_int = field_abs2/I_zero;

end
```

Listing 14.A.4.6
Solves for TE electric field.

```
function TE_mode_field = TE_field(beta,index_mesh,x,k_zero)
% Determines TE optical field for all layers
%
% x - grid created in mesh_x.m
TotalMesh = length(x);   % total number of mesh points
%
zz=beta*k_zero;
%
% Creation of constants at each mesh point
kappa = 0;
for n = 1:(TotalMesh)
   kappa(n)=sqrt((k_zero*index_mesh(n))^2-zz^2);
end
%
% Establish boundary conditions in first layer (substrate).
% Values of the fields U and V are numbered by index not by
% location along x-axis.
% For visualization purposes boundary conditions are set at
% first point.
U(1) = 1.0;
temp = imag(kappa(1));
if(temp<0), kappa(1) = - kappa(1);
end
% The above ensures that we get a field decaying in the substrate
V(1) = kappa(1);
%
for n=2:(TotalMesh)
   cc=cos( kappa(n)*(x(n)-x(n-1)) );
   ss=sin( kappa(n)*(x(n)-x(n-1)) );
   m(1,1)=cc;
   m(1,2)=-1i/kappa(n)*ss;
   m(2,1)=-1i*kappa(n)*ss;
   m(2,2)=cc;
   %
   U(n)=m(1,1)*U(n-1)+m(1,2)*V(n-1);
   V(n)=m(2,1)*U(n-1)+m(2,2)*V(n-1);
```

```
end
%
TE_mode_field = abs(U);                    % Finds Abs(E)
max_value = max(TE_mode_field);
%
% h = plot(x,TE_mode_field/max_value);
% plot normalized value of TE field
% % adds text on x-axix and size of x label
% xlabel('x (microns)','FontSize',22);
% % adds text on y-axix and size of y label
% ylabel('TE electric field','FontSize',22);
% set(h,'LineWidth',1.5);    % new thickness of plotting lines
% set(gca,'FontSize',22);    % new size of tick marks on both axis
% pause
% close all
```

14.6.5 Semiconductors

Listing 14.A.5.1
Determines boundary conditions for potential.

```
function [psi_L, psi_R] = boundary_psi(N,dop)
% File name: boundary_psi.m
% Provides values of potential at contacts without bias
% Ohmic contacts are assumed
% Contacts are numbered as i=1 (left) and i=N (right)
% for method 2

% left contact
a_1 = 0.5*dop(1);
if (a_1>0)
    temp_1  = a_1*(1.0 + sqrt(1.0+1.0/(a_1*a_1)));
    elseif(a_1<0)
        temp_1 = a_1*(1.0 - sqrt(1.0+1.0/(a_1*a_1)));
end
psi_L = log(temp_1);

% right contact
a_N = 0.5*dop(N);
if (a_N>0)
    temp_N  = a_N*(1.0 + sqrt(1.0+1.0/(a_N*a_N)));
    elseif(a_N<0)
        temp_N = a_N*(1.0 - sqrt(1.0+1.0/(a_N*a_N)));
end
psi_R = log(temp_N);
```

Listing 14.A.5.2
Determines electric currents.

```
function tot_curr_sum = currents(N,mesh_layer,psi,n,p)

% File name: currents.m
% Calculates total current due to electrons and holes

mu_n_mesh = mobility_n_AlGaAs_mesh(mesh_layer);
mu_p_mesh = mobility_p_AlGaAs_mesh(mesh_layer);

tot_curr_sum = 0;
for i = 2:N-1
        curr_n(i) = mu_n_mesh(i)*(n(i+1)*ber(psi(i+1)
        - psi(i))...
            - n(i)*ber(psi(i) - psi(i+1)));
            % Electron current density

        curr_p(i) = mu_p_mesh(i)*(p(i)*ber(psi(i+1)-psi(i))...
            - p(i+1)*ber(psi(i)-psi(i+1)));
            % Hole current density

    tot_curr(i) = curr_n(i) + curr_p(i);
    tot_curr_sum(i) = tot_curr_sum(i-1)+ tot_curr(i);
end

end
```

Listing 14.A.5.3
Defines doping at all mesh points.

```
function dop_mesh = doping_mesh(mesh_layer)
%-----------------------------------------------------------------
% File name: doping_mesh.m
% Purpose:   Assigns the value of doping for all mesh points.
%   That value stays the same within each layer and changes
%            from layer to layer.
% dop_mesh        - keeps doping for each mesh point

param_3_layer_AlGaAs_laser;

NumberOfLayers = length(dop_layer);
%
i_mesh = 1;
```

```
for k = 1:NumberOfLayers          % loop over all layers
    for i = 1:mesh_layer(k)          % loop within layer
        dop_mesh(i_mesh) = dop_layer(k);
        i_mesh = i_mesh + 1;
    end
end
```

Listing 14.A.5.4
Determines density of electrons

```
function n = electron_density_2(N,dx,mesh_layer,dop,psi,n,p,
S,I_int)
% File name: electron_density_2.m
% Determines electron density using method 2

dx2 = dx*dx;

mu_mesh = mobility_n_AlGaAs_mesh(mesh_layer);

for i = 2: N-1
        an(i) = mu_mesh(i)*ber(psi(i-1) - psi(i));
        cn(i) = mu_mesh(i)*ber(psi(i+1) - psi(i));
        bn(i) = -(mu_mesh(i)*ber(psi(i) - psi(i-1))
                + mu_mesh(i)*ber(psi(i) - psi(i+1)));
        fn(i) = dx*(recombination_SRH(i,dop,n,p)
                + recombination_stim(i,n,p,S,I_int));
end

an(1) = 0;
cn(1) = 0;
bn(1) = 1;
fn(1) = n(1);
an(N) = 0;
cn(N) = 0;
bn(N) = 1;
fn(N) = n(N);

N = N;
for i=1:N
    A(i,i) = bn(i);             % diagonal elements
end
for i=1:N-1
    A(i,i+1) = cn(i);                % above diagonal
end
for i=1:N-1
```

```
    A(i+1,i) = an(i+1);                    % below diagonal
end

fn=fn';
n = A\fn;                                  % 0.329684 seconds. good solution

end
```

Listing 14.A.5.5
Determines equilibrium properties.

```
function [psi_eq,n,p] = equilibrium_april(N,x_plot,dx,mesh_
layer)
% Function determines equilibrium properties
scaling_dd;
dop = doping_mesh(mesh_layer);      % doping at mesh points
eps_mesh = epsilon_mesh(mesh_layer);

[psi] = initial(N,dop);   % initializes field. No scaling inside

psi_eq = potential_eq_feb(N,dx,psi,dop,eps_mesh);


for i = 2:N-1   % determines equilibrium quantities for plotting
    ro(i) = -n_int*(exp(psi_eq(i))-exp(-psi_eq(i))-dop(i));
    % total charge
    el_field(i) = -(psi_eq(i+1) - psi_eq(i))*V_T/(dx*x_0);
end

for i = 1:N
    n(i) = exp(psi_eq(i));
    p(i) = exp(-psi_eq(i));
end

% Define values at boundaries (for plotting)
el_field(1) = el_field(2);
el_field(N) = el_field(N-1);
ro(1) = ro(2);
ro(N) = ro(N-1);

toc

% redefine equilibrium quantities for plotting
nf1 = n*n_int*1d-6;          % convert to cm^-3
pf1 = p*n_int*1d-6;          % convert to cm^-3
```

```
psi_eq_plot = V_T*psi_eq;      % potential in eV
el_field = el_field*1d-2;       % convert to V/cm
ro_f = q*ro*1d-6;              % convert to C/cm^3

% % %==== Plots of equilibrium values ==========
% figure(1)
% plot(x_plot, psi_eq_plot,'LineWidth',1.5)
% xlabel('x [um]');
% ylabel('Potential [eV]');
% pause
%
% figure(2)
% plot(x_plot, el_field,'LineWidth',1.5)
% xlabel('x [um]');
% ylabel('Electric field [V/cm]');
% pause
%
% figure(3)
% plot(x_plot, ro_f,'LineWidth',1.5)
% xlabel('x [um]');
% ylabel('Total charge density [C/cm^3]');
% pause
%
% close all

end
```

Listing 14.A.5.6
Determines density of holes.

```
function p = hole_density_2(N,dx,mesh_layer,dop,psi,n,p,
S, I_int)
% File name: hole_density_2.m
% Determines hole density using method 2

dx2 = dx*dx;
mu_mesh = mobility_p_AlGaAs_mesh(mesh_layer);

for i = 2: N-1
    ap(i) =  mu_mesh(i)*ber(psi(i) - psi(i-1));
    cp(i) =  mu_mesh(i)*ber(psi(i) - psi(i+1));
    bp(i) = -(mu_mesh(i)*ber(psi(i-1) - psi(i))
            + mu_mesh(i)*ber(psi(i+1) - psi(i)));
    fp(i) = dx*(recombination_SRH(i,dop,n,p)
            + recombination_stim(i,n,p,S,I_int));
```

```
end

ap(1) = 0;
cp(1) = 0;
bp(1) = 1;
fp(1) = p(1);
ap(N) = 0;
cp(N) = 0;
bp(N) = 1;
fp(N) = p(N);

for i=1:N
    A(i,i) = bp(i);              % diagonal elements
end
for i=1:N-1
    A(i,i+1) = cp(i);                   % above diagonal
end
for i=1:N-1
    A(i+1,i) = ap(i+1);                     % below diagonal
end

fp=fp';
p = A\fp;

end
```

Listing 14.A.5.7
Determines initial potential.

```
function [psi] = initial(N,dop)
% Initialize potential based on the requirement of charge
% neutrality throughout the whole structure

for i = 1: N
    zz = 0.5*dop(i);
    if(zz > 0)
        xx = zz*(1 + sqrt(1+1/(zz*zz)));
    elseif(zz <  0)
        xx = zz*(1 - sqrt(1+1/(zz*zz)));
    end
    psi(i) = log(xx);
end

end
```

Listing 14.A.5.8
Determines quantities in non-equilibrium.

```
function [av_curr,n,p] = nonequilibrium_april(N,dx,x_plot,
            mesh_layer,psi,n,p,dop,eps_mesh,Va,S,I_int)
% Solving for the non-equillibirium case
% psi is equilibrium value

scaling_dd;
delta_acc = 1E-7;               % Preset the tolerance

%     k = 1
%Va
        psi(1) = psi(1) + Va;
        flag = 0;
        k_iter = 0;
        while(~flag)    % convergence loop
            k_iter = k_iter + 1;
%           n = electron_density_2(N,dx,mesh_layer,dop,psi,n,p);
            n = electron_density_2(N,dx,mesh_layer,dop,psi,n,
            p,S,I_int);
            p = hole_density_2(N,dx,mesh_layer,dop,psi,n,p,S,I_int);
            [psi, delta_max] = potential_noneq_2(N,dx,dop,psi,n,p,
            eps_mesh);
            if(delta_max < delta_acc)
                flag = 1;
            end
        end             % End of the convergence loop (while loop)
%===============================================================

for i = 2:N-1   % determines electric field in non-equilibrium
    el_field_non_eq(i) = -(psi(i+1) - psi(i))*V_T/(dx*x_0);
end
el_field_non_eq(1) = el_field_non_eq(2);
el_field_non_eq(N) = el_field_non_eq(N-1);

% Calculates currents of electrons and holes

tot_curr_sum = currents(N,mesh_layer,psi,n,p);
av_curr = tot_curr_sum(N - 1)*1e-3/(N-2);

        for i = 2:N-1
            ro_non_eq(i) = - n_int* (n(i) - p(i) - dop(i));
            if(i>1)
            end
        end

        nf = n * n_int*1d-6;
        pf = p * n_int*1d-6;
%         V(k) = Va*V_T;
```

```
%          k = k + 1;
%     end                % end of voltage loop

aa = q*n_int*D_0/(dx*L_Dint*1d2);    % convert to A/cm^2

av_curr = aa*av_curr;

%toc

el_field_non_eq = el_field_non_eq*1d-2;    % convert to V/cm
ro_non_eq = q*ro_non_eq*1d-6;              % convert to C/cm^3

% figure(1);
% semilogy(x_plot,nf1,'LineWidth',1.5);
% Plotting the final carrier densities
% hold on;
% semilogy(x_plot,pf1,'LineWidth',1.5);
% xlabel('x [microns]');
% ylabel('Carrier densities [1/cm^3]');
% pause

% figure(2);
% plot(x_plot,el_field_non_eq,'LineWidth',1.5);
% Plotting the final potential
% xlabel('x [microns]');
% ylabel('Electric field in non-equilibrium [V/cm]');
% pause
%
% figure(3);
% plot(x_plot(1:N-1),ro_non_eq,'LineWidth',1.5);
% Plotting the total charge
% xlabel('x [microns]');
% ylabel('Total charge [C/cm^3]');
% pause
%
% %============= current too large, factor of
% 10 ===== !!!!!!!!!!!!!!!!!!!!!
% figure(4);
% plot(V,av_curr,'LineWidth',1.5);    % Plotting the average current
% xlabel('Voltage [V]');
% ylabel('Current [A/cm^2]');
% pause
%
% close all

end
```

Listing 14.A.5.9

Driver.

```
% File name: pin_hetero_CP_cold.m
% 1D drift-diffusion model of heterogeneous p-n junction
% May 1, 2023
% I do not determine density of photons
% Density of photons is assumed
% Instead, use Eq. 5.17 from Coldren and Corzine book (1995)
% which links
% output power and current and current at threshold


clear all
close all
format long

tic

%'calculating'

scaling_dd;

[N,x_plot,dx,mesh_layer] = mesh_x_CP();
% x in microns, good for plotting

dop = doping_mesh(mesh_layer);        % doping at mesh points
eps_mesh = epsilon_mesh(mesh_layer);

[psi] = initial(N,dop);  % initializes field. No scaling inside

[psi_eq,n,p] = equilibrium_april(N,x_plot,dx,mesh_layer);

%----- add optics here -----------------------
% from laser_test_2.m  -- dobrze liczy
[I_int] = slab_laser(x_plot,mesh_layer);
% determines normalized intensity for all mesh points
                                   % trapz(x,I_int) = 1

%-----------------------------------------------------
Va_max = 0.67;              % max value of applied voltage [V]
dVa = 0.05;                 % value of step of applied voltage [V]
% Scale voltages
Va_max = Va_max/V_T;
dVa = dVa/V_T;

S = 1d-5;
```

```
% First test
for Va = 0:dVa:Va_max
    [av_curr,nf,pf] = nonequilibrium_april(N,dx,x_plot,
    mesh_layer,
            psi_eq,n,p,dop,eps_mesh,Va,S,I_int);
    [S_out1,g3,a3] = rate_eq(x_plot,I_int,nf,pf,S);
end
% Determine threshold values
%----------------------------------------
% Calculations below threshold for assumed value of S
% (density of photons)

k1 = 1;
for Va = 0:dVa:Va_max
    [av_curr1,n1,p1] = nonequilibrium_april(N,dx,x_plot,
    mesh_layer,
            psi_eq,n,p,dop,
        eps_mesh,Va,S,I_int);
    k1 = k1 +1;
    [F_out,g1,a1,tot_den_n,tot_den_p,R_spon_bar]=rate_eq(x_plot,
            I_int,n1,p1,S);
    if g1>a1
        break
    end
end

toc
%'values at threshold'
V_th = Va;
g_th = g1;

%--- Calculations of gain below threshold ---------------
k = 1;
for Va = 0:dVa:V_th
    [av_curr(k),n1,p1] = nonequilibrium_april(N,dx,x_plot,
        mesh_layer,psi_eq,n,p,
        dop,eps_mesh,Va,S,I_int);
    [F_out,g1(k),a1,tot_den_n,tot_den_p(k),R_spon_bar]=
    rate_eq(x_plot,
            I_int,
        n1,p1,S);
    k = k +1;
end
toc
zz = g_th - g1;
```

```
zz;
S_0 = R_spon_bar./zz;
% Output power is proportional to density of photons
tot_den_p = tot_den_p*C_0*1d-6;   % convert to cm^-3
tot_den_p = tot_den_p*1d-18;
plot(tot_den_p,S_0, 'LineWidth',1.5)
xlabel('Carrier density [x 10^{18} cm^{-3}]');
ylabel('Output density of photons [Arbit. units]');
pause
close all
```

Listing 14.A.5.10
Determines potential in equilibrium.

```
function psi = potential_eq_feb(N_mesh,dx,psi,dop,ep)
% File name: solution_psi_2.m
% Performs calculations for p-n junction in equilibrium
% using method 2
% Determines psi (potential) in the iteration process
% using Matlab
% functions

physical_const;        % needed for eps_r
dx2 = dx*dx;

error_iter = 0.0001;
relative_error = 1.0;       % Initial value of the relative error

while(relative_error >= error_iter)
    for i = 1: N_mesh      % POPRAWIC ....................
        a(i) = ep(i);
        c(i) = ep(i);
        b(i) = -(2*ep(i)+(dx2)*(exp(-psi(i))+exp(psi(i))));
        f(i)=(dx2)*(exp(psi(i))-exp(-psi(i))-...
            dop(i)-psi(i)*(exp(psi(i))+exp(-psi(i))));
        end
% Establishing boundary conditions
a(1) = 0; c(1) = 0; b(1) = 1;
a(N_mesh) = 0; c(N_mesh) = 0; b(N_mesh) = 1;
%
% Establishing boundary conditions for potential at contacts
[psi_L, psi_R] = boundary_psi(N_mesh,dop);

f(1) = psi_L;
f(N_mesh) = psi_R;
```

```
%
% Creation of main matrix
for i=1:N_mesh
    A(i,i) = b(i);              % diagonal elements
end
%
for i=1:N_mesh-1
    A(i,i+1) = c(i);           % above diagonal
    A(i+1,i) = a(i+1);          % below diagonal
end

ff=f';

psi_1 = A\ff;
psi_1 = psi_1';

delta_psi = psi_1 - psi;
relative_error = max(abs(delta_psi)./psi_1);

psi = psi_1;

end      % end of while loop

end      % end of function
```

Listing 14.A.5.11
Determines potential in non-equilibrium.

```
function [psi, delta_max] = potential_noneq_2(N,dx,dop,
psi,n,p,ep)
% File name: potential_noneq_2.m
% Solution of the Poisson's equation in non-equibrium
% Using method 2

delta_acc = 1E-7;              % Preset the tolerance

dx2 = dx*dx;
% Define coefficient of main matrix and RHS
for i = 2: N-1
    a(i) = ep(i-1);
    c(i) = ep(i+1);
    b(i) = -(2*ep(i)+dx2*(p(i)+n(i)));
    f(i) = -dx2*(p(i) - n(i) + dop(i) + psi(i)*(p(i) + n(i)));
end
```

```
% Initialize values at contacts
a(1) = 0;
c(1) = 0;
b(1) = 1;
f(1) = psi(1);
a(N) = 0;
c(N) = 0;
b(N) = 1;
f(N) = psi(N);

for i=1:N
    A(i,i) = b(i);              % diagonal elements
end

    A(1,2) = ep(1);
for i=1:N-1
    A(i,i+1) = c(i);            % above diagonal
    A(i+1,i) = a(i+1);          % below diagonal
end

ff=f';

psi_1 = A\ff;
psi_1 = psi_1';

delta = psi_1 - psi;

psi = psi_1;

% Test update in the outer iteration loop

delta_max = 0;
for i = 1: N
    xx = abs(delta(i));
    if(xx > delta_max)
        delta_max=xx;
    end
end

end
```

Bibliography

[1] S. Seki, T. Yamanaka, and K. Yokoyama. Two-dimensional numerical analysis of current blocking mechanism in InP buried heterostructure lasers. *J. Appl. Phys.*, 71:3572–3578, 1992.

[2] G.-L. Tan, N. Bewtra, K. Lee, and J.M. Xu. A two-dimensional non-isothermal finite element simulation of laser diodes. *IEEE J. Quantum Electron.*, 29:822–835, 1993.

[3] K.J. Ebeling. *Integrated Optoelectronics. Waveguide Optics, Photonics, Semiconductors.* Springer-Verlag, Berlin, 1993.

[4] Z.-M. Li, K.M. Dzurko, A. Delage, and S.P. McAlister. A self-consistent two-dimenional model of quantum-well semiconductor lasers: optimization of a GRIN-SCH SQW laser structure. *IEEE J. Quantum Electron.*, 28:792–803, 1992.

[5] T. Ohtoshi, K. Yamaguchi, C. Nagaoka, T. Uda, Y. Murayama, and N. Chinone. A two-dimensional device simulator of semiconductor lasers. *Solid State Electron.*, 30:627–638, 1987.

[6] B.W. Hakki. Instabilities in output of injection lasers. *J. Appl. Phys.*, 50:5630–5637, 1979.

[7] M. Gault, P. Mawby, A.R. Adams, and M. Towers. Two-dimensional simulation of constricted-mesa InGaAsP/InP burried-heterostructure lasers. *IEEE J. Quantum Electron.*, 30:1691–1700, 1994.

[8] L.A. Coldren and S.W. Corzine. *Diode Lasers and Photonic Integrated Circuits.* Wiley, New York, 1995.

[9] M. Shur. *Physics of Semiconductor Devices.* Prentice Hall, Englewood Cliffs, NJ, 1990.

[10] S.J. Polak, C. Den Heijer, and W.H.A. Schilders. Semiconductor device modelling from the numerical point of view. *International Journal for Numerical Methods in Engineering*, 24:763–838, 1987.

15

Conclusions

In this work I concentrated on the development of main elements of the Matlab software. With it the reader can perform extensive simulations and also extend if needed.

Main examples were performed on AlGaAs material system. However, with all information provided one can easily simulate other systems, like InP based.

After finalizing main text I can suggest possible future extensions.

- develop 2D model

- implement Newton method

- implement models of transport across heterointerface

- add more model of losses

- include more complicated gain models

- analyze single and multiple quantum well systems.

As a possible justification of one of the above suggestions, I want to mention work by Grupen, Hess and Song [1]. They reported different approaches to transport across heterointerface which resulted in significant differences in I-V characteristics of forward biased junction.

Also, as indicated by several groups, see e.g. [2] the common decoupled Gummel scheme becomes unsuitable at large values of injection currents. In such situations there exists strong coupling between equations and one should use Newton-Raphson method.

When (and if) I get more time (and energy) I plan to produce the Second Edition of this book and extend it by expanding some of the above topics. Especially, Newton method should allow to analyze structures for larger photon densities, i.e. larger power values and determine density of photons in a self-consistent way.

Bibliography

[1] M. Grupen, K. Hess, and G.H. Song. Simulation of transport over hetero-junctions. In W. Fichtner and D. Aemmer, editors, *Simulation of Semiconductor Devices and Processes*, volume 4, pages 303–311. Hartung-Gorre Verlag, Konstanz, 1991.

[2] M. Gault, P. Mawby, A.R. Adams, and M. Towers. Two-dimensional simulation of constricted-mesa InGaAsP/InP burried-heterostructure lasers. *IEEE J. Quantum Electron.*, 30:1691–1700, 1994.

A

Material Parameters

We start with a summary of general materials and in the following sections provide details of AlGaAs and InGaAsP.

A.1 Some Properties of Important Materials

First, we summarize main properties of some general class of materials.

A.1.1 Bandgap energies

Formulas used to determine the bandgap energies of some materials for an arbitrary compositions x and y at room temperature ($T = 300K$) are summarized in a Table A.1 [1].

A.1.2 Mobilities

At around room temperature and at low electric fields the temperature dependence of mobility is approximated as [2]

$$\mu(T) = \mu_0 \left(\frac{T}{300} \right)^{\delta_0} \tag{A.1}$$

Doping dependence on mobility is

$$\mu\left(N_{dop}\right) = \mu_{dop} + \frac{\mu_0 - \mu_{dop}}{1 + \left(\frac{N_{dop}}{N_{ref}}\right)^{\alpha}} \tag{A.2}$$

where μ_0 is the maximum mobility (no doping) and μ_{dop} is the minimum mobility for high doping densities. Several of the above parameters are summarized in Table A.2, from [2]. For small values of electric fields mobility is almost constant.

DOI: 10.1201/9781003265849-A

TABLE A.1

Values of bandgap energies for for some semiconductor
materials

Material	Bandgap energy
$In_{1-x}Ga_xAs$	$E_g(x) = 0.35 + 0.629x + 0.436x^2$
$InAs_yP_{1-y}$	$E_g(y) = 0.35 - 1.17y + 0.18y^2$
$In_{1-x}Ga_xP$	$E_g(x) = 1.35 + 0.668x + 0.758x^2$
$GaAs_yP_{1-y}$	$E_g(y) = 2.77 - 1.56y + 0.21y^2$
$In_{1-x}Ga_xAs_yP_{1-y}$	$E_g(x,y) = 0.35 + 0.668x - 0.17y$
	$\qquad +0.758x^2 + 0.18y^2 - 0.069xy$
	$\qquad -0.322x^2y + 0.03xy^2$

TABLE A.2

Mobility parameters for some semiconductor materials

Symbol	$\mu_0 [cm^2/Vs]$	$\mu_{dop}[cm^2/Vs]$	$N_{ref}[10^{17}cm^{-3}]$	α	δ_0
Si	1430	80	1.12	0.72	02.0
Ge	3895	641	0.613	1.04	−1.67
GaAs	9400	500	0.6	0.394	−2.1
InP	5200	400	3.0	0.42	−2.0
GaP	152	10	44.0	0.80	−1.6
AlAs	400	10	5.46	1.0	−2.1

The electron and hole velocities at large values of electric field are approximated as [3]

$$v_n = \frac{\mu_n F}{\left[1 + \left(\frac{\mu_n F}{v_s}\right)^2\right]^{1/2}} \tag{A.3}$$

and

$$v_p = \frac{\mu_p F}{1 + \left(\frac{\mu_p F}{v_s}\right)} \tag{A.4}$$

where (for electrons)

$$\mu_n = \mu_{mn} + \frac{\mu_{on}}{1 + \frac{N_T}{N_{cn}}\nu} \tag{A.5}$$

$$\mu_{mn} = 88 \left(\frac{T}{300}\right)^{-0.57} \left(\frac{cm^2}{Vs}\right) \tag{A.6}$$

$$\mu_{on} = 7.4 \times 10^8 \ T^{-2.33} \ \left(\frac{cm^2}{Vs}\right) \tag{A.7}$$

$$\nu = 0.88 \left(\frac{T}{300}\right)^{-0.146} \tag{A.8}$$

$$N_{cn} = 1.26 \times 10^{17} \left(\frac{T}{300}\right)^{2.4} \ (cm^{-3}) \tag{A.9}$$

and (for holes)

$$\mu_{mp} = 54 \left(\frac{T}{300}\right)^{-0.57} \quad \left(\frac{cm^2}{Vs}\right) \tag{A.10}$$

$$\mu_{op} = 1.36 \times 10^8 \; T^{-2.33} \quad \left(\frac{cm^2}{Vs}\right) \tag{A.11}$$

$$N_{cp} = 2.35 \times 10^{17} \left(\frac{T}{300}\right)^{2.4} \quad \left(cm^{-3}\right) \tag{A.12}$$

The saturation velocity v_s for electrons and holes is

$$v_s = \frac{2.4 \times 10^5}{1 + 0.8 \exp\left(\frac{T}{600}\right)} \quad \left(\frac{m}{s}\right) \tag{A.13}$$

Exponent ν is the same for electrons and holes.

A.2 Practical Material: $Al_xGa_{1-x}As$

In practical applications important role is played by two material systems: $Al_xGa_{1-x}As$ and $In_{1-x}Ga_xAs_yP_{1-y}$. They form basis to fabricate devices which operate at different wavelengths. There are also other important materials, like those based on GaN. However, in this book we only consider devices based on GaAs. Additionally, for future developments we also summarize here properties of InP based materials.

A.2.1 Band structure parameters

The parameters are listed in Table A.3. The values and formulas were taken from [4].

A.2.2 Band discontinuity

Here we summarize the so-called band-offset problem. It is associated with band discontinuity at heterointerface. The band discontinuity at an abrupt heterostructure is modeled by the Anderson electronic affinity rule [5] (compare Fig. 2.12).

There is a considerable amount of experimental data on band offsets. For $Al_xGa_{1-x}As/GaAs$ material system one has the following result [6]

$$E_g(GaAs) = 1.424 \quad eV \; (300K) \tag{A.14}$$

$$E_g(Al_xGa_{1-x}As) = 1.424 + 1.247x \quad eV \; (300K) \tag{A.15}$$

$$\Delta E_g(x) = E_g(Al_xGa_{1-x}As) - E_g(GaAs) = 1.247x \quad eV$$

TABLE A.3
Formulas for band structure parameters for GaAs-AlGaAs; x denotes Al concentration.

Parameter	Dependence on the mole fraction x
Energy bandgap	$E_g = 1.424 + 1.247x \ (x < 0.45)$ $1.9 + 0.125x + 0.143x^2 (x > 0.45)$
Electron affinity	$\chi_e = 4.70 - 0.748x \ (x < 0.45)$ $\chi_e = 3.80 - 0.14x$
Effective mass for electrons	$m_n/m_0 = 0.067 + 0.083x \ (x < 0.45)$ $m_n/m_0 = 0.85 - 0.14x \ (x > 0.45)$
Effective mass for holes	$m_{lh} = (0.087 + 0.063x)\, m_0$ $m_{hh} = (0.62 + 0.14x)\, m_0$ $m_p = (m_{lh}^{3/2} + m_{hh}^{3/2})^{2/3}$

$$\Delta E_c(x) = 0.67 \Delta E_g(x) \tag{A.16}$$

$$\Delta E_v(x) = 0.33 \Delta E_g(x) \tag{A.17}$$

The electronic affinity is adjusted to measured values of the conduction band discontinuity ΔE_c (Fig. 2.12).

$$\Delta \chi_e = \Delta E_c \tag{A.18}$$

A.2.3 Doping

For *AlGaAs/GaAs* heterojunction simulations, the incomplete ionization of impurities is taken into account following [7] as follows

$$N_D^+ = \frac{N_D}{1 + \frac{n}{N_c} g_D \, \exp(\frac{\Delta E_D}{k_B T})} \tag{A.19}$$

$$N_A^- = \frac{N_A}{1 + \frac{p}{N_v} g_A \, \exp(\frac{\Delta E_A}{k_B T})} \tag{A.20}$$

where N_D are N_A donor and acceptor impurity concentrations, g_D and g_A are the ground state degeneracy factors of the donor and acceptor levels, ΔE_D and ΔE_A are the donor and acceptor activation energies and N_c, N_v are effective densities of states in the conduction and valence bands, respectively.

In the above formulas, it is assumed that the shallow impurities are in equilibrium with the local carriers, i.e. they share the same quasi-Fermi levels. The ionisation from deep level traps is ignored. The values of doping parameters

TABLE A.4
Doping parameters for AlGaAs.

Symbol	Value	Description
N_D	$5 \times 10^{17} cm^{-3}$	Donor impurity concentration
N_A	$1 \times 10^{18} cm^{-3}$	Acceptor impurity concentration
g_D	2	Ground state degeneracy factor for donors
g_A	4	Ground state degeneracy factor for acceptors
E_D	$0.005 eV$	Activation energy for donor
E_A	$0.026 eV$	Activation energy for acceptor
N_{res}	$2 \times 10^{15} cm^{-3}$	Residual concentration

for AlGaAs are summarized in Table A.4. They were compiled from Young et al [7].

A.2.4 Carrier mobilities

There are many models carrier's mobilities. The model used here applies to $Al_xGa_{1-x}As$ material system. The low-field mobilities for the electrons and holes are expressed as follows [7]

$$\mu_n = f_n(x) \frac{7200}{[1 + 5.51 \times 10^{-17}(N_D + N_A)]^{0.233}} \times (\frac{300}{T})^{2.3} cm^2/Vs \quad (A.21)$$

$$\mu_p = f_p(x) \frac{380}{[1 + 3.17 \times 10^{-17}(N_D + N_A)]^{0.266}} \times (\frac{300}{T})^{2.7} cm^2/Vs \quad (A.22)$$

where x is the Al composition. The electric field dependence of mobilities is not considered for simplification. The functions $f_n(x)$ and $f_p(x)$ model compositional dependence of the mobilities with $f(0) = 1$. Their forms are

$$f_n(x) = 1 - 1.27x \quad (A.23)$$

$$f_p(x) = 1 - 0.67x \quad (A.24)$$

Functions $f_n(x)$ and $f_p(x)$ introduced in [7] account for the reduction of the original mobilities of GaAs due to Al presence. In the above N_D and N_A are the concentrations of donors and acceptor, respectively.

A.2.5 Optical parameters

Optical parameters for AlGaAs used in the model described here are summarized in Table A.5, see [4] for values of $\bar{n}$ and ε_{dc}. Gain peak dependence

TABLE A.5
Optical parameters and their formulas for AlGaAs used in the model; x refers to Al composition.

Physical Quantity	Symbol	Formula or value	Units
Real refractive index	$\bar{n}$	$3.65 - 0.73x$	
Dielectric constant	ε_{dc}	$13.1 - 3x$	
Local gain constant	a	2.4×10^{-16}	cm^2
Transparency density	n_t	2.1×10^{18}	cm^{-3}

is modeled after [8] and [9]. We assume linear dependence with the carrier density as follows

$$g = a(min(n, p) - n_t) \tag{A.25}$$

with coefficients summarized in Table A.5, after Ebeling [8].

A.2.6 Recombination parameters

Values of all the above recombination parameters are summarized in Table A.6. τ_{p0} and τ_{n0} are taken from ref. [7].

A.2.7 Losses

We included the following physical mechanisms responsible for absorption (losses): free-carrier absorption, cavity absorption, mirror losses. Intervalence band absorption is neglected for AlGaAs system.

The numerical values of absorption coefficients are summarized in Table A.7. They are based on work by Li et al [4] and [10].

Mirror loss α_m depends on facet reflectivity R. Here, we assumed identical facets and used the value of $R = 0.31$.

TABLE A.6
Recombination coefficients and their numerical values for AlGaAs.

Physical Quantity	Symbol	Value
Spontaneous recombination coefficient	B_0	$2 \times 10^{-16} m^3 s^{-1}$
Auger recombination coefficients	C_n	$2.1 \times 10^{-42} m^6 s^{-1}$
	C_p	$2.1 \times 10^{-42} m^6 s^{-1}$
Carrier lifetimes	τ_{p0}	$10^{-9} s$
	τ_{n0}	$10^{-9} s$

TABLE A.7
Absorption coefficients and their numerical values
for AlGaAs.

Physical Quantity	Symbol	Value
Free-carrier absorption	α_{fn}	$3 \times 10^{-22} m^2$
	α_{fp}	$7 \times 10^{-22} m^2$
Mirror reflectivity	R	0.31

A.3 $In_{1-x}Ga_xAs_yP_{1-y}$ Material System

Models presented assume low doping concentration in the active layers (below $5 \times 10^{17} cm^{-3}$). The models may be inaccurate for higher doping concentrations.

A.3.0.1 Band structure parameters

For $In_{1-x}Ga_xAs_yP_{1-y}$ system lattice-matched to InP, the parameters associated with the band structure are summarized in the Table A.8 (from [6] Table K.3) (m_0 is the mass of bare electron)

A.3.1 Band discontinuity

For $In_{1-x}Ga_xAs_yP_{1-y}/InP$ quaternary semiconductor lattice matched to InP substrate one has the following relations [6]

$$x = \frac{0.1896y}{0.4176 - 0.0125y} \tag{A.26}$$

$$E_g(In_{1-x}Ga_xAs_yP_{1-y}) = 1.35 - 0.774y + 0.149y^2 \ eV \tag{A.27}$$

$$\Delta E_g(y) = 0.775y - 0.149y^2 \ eV \tag{A.28}$$

$$\Delta E_v(y) = 0.502y - 0.152y^2 \ eV \tag{A.29}$$

TABLE A.8
The values of parameters for InGaAsP associated with band structure.

Parameter	Dependence on the mole fraction y
Energy gap at zero doping	$E_g(y) = 13.5 - 0.775y + 0.149y^2 \ (eV) at 298K$
Conduction-band effective mass	$m_c = (0.080 - 0.039) m_0$
Heavy-hole effective mass	$m_{hh} = 0.46m_0$
Light-hole effective mass	$m_{lh} = (0.12 - 0.099y + 0.030y^2) m_0$

$$\Delta E_c(y) = \Delta E_g(y) - \Delta E_v(y) = 0.273y + 0.003y^2 \ eV \qquad (A.30)$$

There are also other usefull relations. The electronic affinity is adjusted to measured values of the conduction band discontinuity ΔE_c. The relation due by Forrest et al [11] is used

$$\Delta E_c(y) = 0.4\Delta E_g(y) \qquad (A.31)$$

where

$$\Delta E_g(y) = E_g(InP) - E_g(In_{1-x}Ga_xAs_yP_{1-y}) \qquad (A.32)$$

One finds

$$\chi_e(y) = \chi_e(InP) + 0.4\Delta E_g(y) \qquad (A.33)$$

where [12] $\chi_e(InP) = 4.40eV$ is the electronic affinity of InP. The difference in band-gap energies is

$$\Delta E_g(y) = -0.72y + 0.12y^2 \qquad (A.34)$$

and does not depend on temperature.

A.3.2 Doping

For simulations based on $InGaAsP/InP$ material system doping distribution is typically provided by a user. Below we briefly describe examples of modulation doping (MD).

Belenky et al [13] reported on the role of p-doping profile and regrowth on the static characteristics of $1.3\mu m$ multiple quantum well (MQE) $InGaAsP/InP$ lasers using both experimental and theoretical analysis. Their analysis shows that details of the p-doping profile have a systematic impact on the static device performance. In particular, there was an optimal placement of the p-i junction which minimized threshold current and maximized external efficiency.

Zhang et al [14] fabricated and analyzed $1.5\mu m$ n-type $InGaAsP/InGaAsP$ modulation-doped multiple quantum well DFB lasers. They observed the reduction of threshold current with the increased sheet carrier density. Their structure showed $17GHz$ 3 dB modulation bandwidth.

A.3.3 Carrier mobilities

The electron and hole mobilities are given by the following formulas

$$\mu_n(y,T,N) = f_n(y)\frac{5000}{[1+\frac{N_D+N_A}{1.22\times10^{16}}]^{0.191}} \times (\frac{300}{T})^{1/2}cm^2/Vs \qquad (A.35)$$

$$\mu_p(y,T,N) = f_p(y)\frac{192}{[1+\frac{N_D+N_A}{2.71\times10^{16}}]^{0.272}} \times (\frac{300}{T})^{1/2}cm^2/Vs \qquad (A.36)$$

In the above equations, $N = N_D + N_A$ is the total doping concentration expressed in cm^{-3}. The functions $f_n(x)$ and $f_p(x)$ model the compositional dependence of the mobilities and are

$$f_n(y) = 3.04y^2 - 1.30y + 1 \qquad \text{(A.37)}$$

$$f_p(y) = 2.58y^2 - 2.04y + 1 \qquad \text{(A.38)}$$

The $T^{-1/2}$ temperature dependence is a good approximation only for temperatures $T > 200K$. The above formulas are given in [2, 15–18].

A.3.4 Optical parameters

Refractive index (at a wavelength corresponding to the band gap) of $In_{1-x}Ga_xAs_yP_{1-y}$ lattice-matched to InP (for which $x \approx 0.45y$) is [19]

$$\bar{n}_0(y) = 3.4 + 0.256y - 0.095y^2 \qquad \text{(A.39)}$$

Another expression for static dielectric constant ε_{dc} is presented by [20] for $In_{1-x}Ga_xAs_yP_{1-y}$ lattice-matched to InP and is

$$\varepsilon_{dc} = \varepsilon_0(12.40 + 1.5y) \qquad \text{(A.40)}$$

where ε_0 is the free space permittivity.

The bulk index of refraction is calculated using the following formulas [21, 22]

$$\bar{n}_0(\lambda) = \left(A + \frac{B\lambda^2}{\lambda^2 - C} \right)^{1/2}$$

where λ is the emitted wavelength in μm. The parameters A, B, C are

$$
\begin{aligned}
A &= 7.2550 + 1.150y + 0.489y^2 \\
B &= 2.3160 + 0.604y - 0.493y^2 \\
C &= 0.3922 + 0.396y + 0.158y^2
\end{aligned}
$$

A.3.5 Optical gain

The parameters which appear in linear gain model are given by [15]

$$a \equiv a(y, T) = 1.0 \times 10^{-16} \exp\left[\frac{E_g(y, T = 300)}{3.4} \right] \exp\left(\frac{173.8}{T} \right) cm^2 \qquad \text{(A.41)}$$

$$b \equiv b(y, T) = 158.5 \exp\left[\frac{E_g(y, T = 300)}{1.34} \right] \exp\left(-\frac{231.8}{T} \right) cm^{-1} \qquad \text{(A.42)}$$

The temperature dependence is based on Yano's work [23, 24]. Specific values of a and b coefficients at temperature $T = 300$ are

$$a(y = 0.6, T = 300) = 1.35 \times 10^{-16} cm^2$$
$$b(y = 0.6, T = 300) = 150 cm^{-1}$$

A.3.6   Recombination coefficients

Spontaneous recombination rate

The spontaneous recombination rate is given by

$$R_{spon} = B(n, p, T) \cdot (n \cdot p - n_0 p_0) \tag{A.43}$$

where B is the spontaneous recombination coefficient [6] and [2] and n_0 and p_0 are the electron and hole concentrations at thermal equilibrium

$$n_0 \cdot p_0 = n_{int}^2 \tag{A.44}$$

The coefficient $B(n, p, T)$ is given by the formula [15, 25, 26]

$$B(n, p, T) = B_0(T) - B_1(T) \min(n, p) \tag{A.45}$$

where coefficients $B_0(T)$ and $B_1(T)$ have the following values

$$B_0(T) = 8.0 \times 10^{-11} \left(\frac{300}{T} \right)^{3/2} cm^3/s \tag{A.46}$$

$$B_1(T) = 2.0 \times 10^{-19} B_0(T) cm^6/s \tag{A.47}$$

The variation of B with material composition is neglected [15].

Auger recombination

The Auger recombination rate is given by formula

$$R_{Aug} = c_n n (np - n_0 p_0) + c_p p (np - n_0 p_0) \tag{A.48}$$

where c_n and c_p are the Auger recombination coefficients responsible for CHCC and CHSH processes, respectively. Both coefficients are modeled as

$$c_n(y, T) = \left[1.5 \times 10^{-29} \exp \left(\frac{T - 300}{600} \right) \right] \left(\frac{y}{0.6} \right)^3 cm^6/s \tag{A.49}$$

$$c_p(y, T) = \left[1.05 \times 10^{-29} \left(\frac{T}{300} \right)^{3.6} + 0.45 \times 10^{-29} \right] \left(\frac{y}{0.6} \right)^3 cm^6/s \tag{A.50}$$

For InGaAsP at $\lambda = 1.3 \mu m$ and $T = 300 K$, the above equations give

$$c_n(y = 0.6, T = 300) = c_p(y = 0.6, T = 300) = 1.5 \times 10^{-29} cm^6/s \tag{A.51}$$

The non-radiative Shockley-Read-Hall recombination

The non-radiative Shockley-Read-Hall (SRH) recombination rate is given by

$$R_{SRH} = \frac{n \cdot p - n_{int}p_{int}}{\tau_n (n + n_{int}) + \tau_p (p + p_{int})} \tag{A.52}$$

where τ_n and τ_p are carrier lifetimes. Lifetimes depend on the quality of the device. Typical values are in the $1 - 20ns$ range (typically $10ns$).

A.3.7 Absorption coefficients

The free-carrier absorption is given by the general formula as

$$\alpha_{fc}(n,p) = \alpha_{fn}n + \alpha_{fp}p$$

The parameters α_{fn} and α_{fp} are assumed to be constant and have the values [25]

$$\alpha_{fn} = 3.0 \times 10^{-18}cm^2$$
$$\alpha_{fp} = 7.0 \times 10^{-18}cm^2$$

The intervalence band absorption coefficient is more complicated to determine. Here, we adopted the following expression [15]

$$\alpha_{ivba}(p) = \begin{cases} 0 & for \quad T < 200K \\ \kappa_{p1}(y,T)p & for \quad 200K < T < 300K \\ \kappa_{p1}(y)p & for \quad T > 300K \end{cases}$$

where

$$\kappa_{p1}(y,T) = \left(\frac{y}{0.6}\right)^7 \left(-1.485 \times 10^{-17} + 8.0 \times 10^{-20}T\right) cm^2$$

and

$$\kappa_{p1}(y) = \left(\frac{y}{0.6}\right)^7 9.15 \times 10^{-18}cm^2$$

Here, y is the mole fraction in the active region and hole density p is expressed in cm^{-3}. For passive layers, we assume that α_{ivba} is equal to half the absorption coefficient in the active layer. For structures with multiple active layers we must be careful using this approximation.

Cavity absorption coefficient α_{abs} is responsible for absorption due to cavity imperfections and should be controlled by the user. For good quality devices, $\alpha_{abs} = 0.0$. Otherwise, we suggest to take

$$\alpha_{abs} \simeq 10cm^{-1} \tag{A.53}$$

Mirror loss α_m is given by

$$\alpha_m = \frac{1}{L_{cavity}} \ln\left(\frac{1}{R}\right) \tag{A.54}$$

where L_{cavity} is the cavity length and R is the facet reflectivity.

A.3.8   Spontaneous emission factor

Typical value of spontaneous emission factor is assumed to be $\beta = 1 \times 10^{-4}$. This corresponds to an index-guided laser [27]. For a gain-guided laser $\beta = \simeq 15 \times 10^{-3}$ is suggested [27, 28].

A.3.9   Summary of parameters for InP systems

For a quick, practical implementation, in the following two Tables we summarized physical parameters for InGaAsP-InP lasers [15]. Data for the emission wavelength of $1.2\mu m$ and temperature $T = 300K$ are provided in Table A.9; in Table A.10 there are parameters for the emission wavelength of $1.3\mu m$ and temperature $T = 300$.

TABLE A.9
Summary of parameters for InGaAsP-InP at 1.2μ at $T = 300$.

Parameters	Description	InP	InGaAsP
ε_S	static permittivity	12.40	13.12
$\overline{n}_0$	refraction index	3.23	3.50
E_g	bandgap (eV)	1.35	1.03
χ_e	electron affinity (eV)	4.40	4.56
a	gain parameter $(10^{-16}cm^2)$	0	1.32
b	gain parameter (cm^{-1})	0	158
α_a	absorption coefficient due to imperfections (cm^{-1})	10	10
α_n	absorption coefficient due to free carriers (electrons) $(10^{-18}cm^2)$	3.0	3.0
α_p	absorption coefficient due to free carriers (holes) $(10^{-18}cm^2)$	7.0	7.0
κ_p	IVBA coefficient $(10^{-18}cm^2)$	0.7	1.5
c_n	Auger recombination coefficient CHCC process $(10^{-29}cm^6/s)$	0	0.75
c_p	Auger recombination coefficient CHSC process $(10^{-29}cm^6/s)$	0	0.75
B_0	spontaneous recombination coefficient $(10^{-11}cm^3/s)$	8.0	8.0
B_1	recombination coefficient $(10^{-29}cm^6/s)$	1.6	1.6
τ_n, τ_p	non-radiative recombination (SRH) $10^{-9}s$	10	10

TABLE A.10
Summary of parameters for InGaAsP-InP at 1.3μ at $T = 300$.

Parameters	Description	InP	InGaAsP
ε_S	static permittivity	12.40	13.30
$\overline{n}_0$	refraction index	3.21	3.52
E_g	bandgap (eV)	1.35	0.96
χ_e	electron affinity (eV)	4.40	4.56
a	gain parameter $(10^{-16}cm^2)$	0	1.35
b	gain parameter (cm^{-1})	0	150
α_a	absorption coefficient due to imperfections (cm^{-1})	10	10
α_n	absorption coefficient due to free carriers (electrons) $(10^{-18}cm^2)$	3.0	3.0
α_p	absorption coefficient due to free carriers (holes) $(10^{-18}cm^2)$	7.0	7.0
κ_p	IVBA coefficient $(10^{-18}cm^2)$	4.6	9.2
c_n	Auger recombination coefficient CHCC process $(10^{-29}cm^6/s)$	0	1.5
c_p	Auger recombination coefficient CHSC process $(10^{-29}cm^6/s)$	0	1.5
B_0	spontaneous recombination coefficient $(10^{-11}cm^3/s)$	8.0	8.0
B_1	recombination coefficient $(10^{-29}cm^6/s)$	1.6	1.6
τ_n, τ_p	non-radiative recombination (SRH) $10^{-9}s$	10	10

Bibliography

[1] B. Mroziewicz, M. Bugajski, and W. Nakwaski. *Physics of Semiconductor Lasers*. Polish Scientific Publishers, Warszawa, 1991.

[2] J. Piprek. *Semiconductor Optoelectronic Devices. Introduction to Physics and Simulations*. Academic Press, Amsterdam, 2003.

[3] M. Shur. *Physics of Semiconductor Devices*. Prentice Hall, Englewood Cliffs, NJ, 1990.

[4] Z.-M. Li, K.M. Dzurko, A. Delage, and S.P. McAlister. A self-consistent two-dimenional model of quantum-well semiconductor lasers: optimization of a GRIN-SCH SQW laser structure. *IEEE J. Quantum Electron.*, 28:792–803, 1992.

[5] R.L. Anderson. Experiments on Ge-GaAs heterojunctions. *Solid State Electron.*, 5:341–351, 1962.

[6] S.-L. Chuang. *Physics of Photonics Devices.* Wiley, New York, 2009.

[7] K. Yang, J. R. East, and G.I. Haddad. Numerical modeling of abrupt heterojunctions using a thermionic-field emission boundary condition. *Solid State Electron.*, 36:321–330, 1993.

[8] K.J. Ebeling. *Integrated Optoelectronics. Waveguide Optics, Photonics, Semiconductors.* Springer-Verlag, Berlin, 1993.

[9] G.-L. Tan, N. Bewtra, K. Lee, and J.M. Xu. A two-dimensional non-isothermal finite element simulation of laser diodes. *IEEE J. Quantum Electron.*, 29:822–835, 1993.

[10] W.T. Tsang. Extremely low threshold (AlGa)As graded-index waveguide separate-confinement heterostructure lasers grown by molecular beam epitaxy. *Appl. Phys. Let.*, 40:217, 1982.

[11] S.R. Forrest, P.H. Schmidt, R.B. Wilson, and M.L. Kaplan. Relationship between the conduction-band discontinuities and band-gap differences of InGaAsP/InP heterojunctions. *Appl. Phys. Let.*, 45:1199–1201, 1984.

[12] H.C. Casey and M.B. Panish. *Heterostructure Lasers.* Academic Press, New York, 1978.

[13] G.L. Belenky, C.L. Reynolds, D.V. Donetsky, G.E. Shtengel, M.S. Hybertsen, M.A. Alam, G.A. Baraff, R.K. Smith, R.F. Kazarinov, J. Winn, and L.E. Smith. Role of p-doping profile and regrowth on the static characteristics of $1.3 - \mu m$ MQW InGaAsP-InP lasers: experiment and modeling. *IEEE J. Quantum Electron.*, 35:1515–1520, 1999.

[14] R. Zhang, W. Wang, F. Zhou, B. Wang, L. Wang, J. Bian, L. Zhao, H. Zhu, and S. Jian. A $1.5\mu m$ n-type ingaasp/ingaasp modulation-doped multiple quantum well dfb laser by mocvd. *Semicond. Sci. Technol.*, 21:306–310, 2006.

[15] A. Champagne. *Modelisation des lasers InGaAsP-InP a double heterostructure et a double region active.* PhD thesis, Departement de Genie Physique, Ecole Polytechnique, Universite de Montreal, 1992.

[16] S. Asada, S. Sugou, K.-I. Kasahara, and S. Kumashiro. Analysis of leakage current in buried heterostructure lasers with semiinsulating blocking layers. *IEEE J. Quantum Electron.*, 25:1362–1368, 1989.

[17] M.-C. Amann and W. Thulke. Current confinement and leakage currents in planar buried-ridge-structure laser diodes on n-substrate. *IEEE J. Quantum Electron.*, 25:1595–1602, 1989.

[18] T. Ohtoshi, K. Yamaguchi, and N. Chinone. Analysis of current leakage in InGaAsP/InP buried heterostructure lasers. *IEEE J. Quantum Electron.*, 25:1369–1375, 1989.

[19] G.P. Agrawal and N.K. Dutta. *Semiconductor Lasers. Second Edition*. Kluwer Academic Publishers, Boston, 2000.

[20] S. Adachi. Material parameters of $In_{1-x}Ga_xAs_yP_{1-y}$ and related binaries. 53:8775–8792, 1982.

[21] F. Fiedler and A. Schlachetzki. Optical parameters of InP-based waveguides. *Solid State Electron.*, 30:73–83, 1987.

[22] G.D. Pettit and W.J. Turner. Refractive index of inp. *J. Appl. Phys.*, 36:2081, 1965.

[23] M. Yano, H. Imai, and M. Takusagawa. Analysis of electrical, threshold and temperature characteristics of InGaAsP/InP double heterojunction lasers. *IEEE J. Quantum Electron.*, 17:1954–1963, 1981.

[24] M. Yano, H. Imai, and M. Takusagawa. Analysis of threshold temperature characteristics for ingaasp/inp double-heterostructure lasers. *J. Appl. Phys.*, 52:3172–3175, 1981.

[25] A. Champagne, R. Maciejko, and J.M. Glinski. The performance of double active region InGaAsP lasers. *IEEE J. Quantum Electron.*, 27:2238–2247, 1991.

[26] R. Olshansky, C.B. Su, J. Manning, and W. Powazinik. Measurement of radiative and nonradiative recombination rates in InGaAsP and AlGaAs light sources. *IEEE J. Quantum Electron.*, 20:838–854, 1984.

[27] T.R. Chen, M. Kajanto, Y.H. Zhuang, and A. Yariv. Double active region index-guided semiconductor laser. *Appl. Phys. Let.*, 54:108–110, 1989.

[28] K. Peterman. Calculated spontaneous emission factor for double-heterostructure injection lasers with gain-induced waveguiding. *IEEE J. Quantum Electron.*, 15:566–570, 1981.

B

Short History of Semiconductor Laser Simulations

The history of the birth of the semiconductor laser diode has been described by Hecht [1]. It is very interesting historical account with the information that the first step in the development of semiconductor laser was done in 1907 by Henry J. Round working for the Marconi Company who observed yellow light from semiconductor junction. Those were the first observations of light coming from LED (light-emitting diode).

Due to its practical importance laser simulators have been in development by different groups in the last 25 years or so. Here, we will attempt to briefly summarize main developments as described in open literature. We do not pretend to be complete. The description is classified by the company/university where the developments took place. In this part. we do not attempt to provide detailed description of the models and techniques developed in each paper. We only describe main developments and main conclusions. More details about models are discussed in later chapters. In this short review we concentrate on approaches based on drift-diffusion models.

Pre-1984 Attempts

Extensive review of early models and simulation results prior to 1984 have been published by Buus [2]. We will not attempt to duplicate his excellent work. Some of the papers reviewed by Buus formed the basis for later development as discussed below. Those early attempts mostly concentrated on one-dimensional and quasi two-dimensional models.

Main characteristics of existing laser models were summarized in his Table 3. Different basic assumptions were used by different authors to construct specific models; this means that their validity depends strongly on the specific laser structure to which the model applies.

DOI: 10.1201/9781003265849-B

For more information, the reader should consult Buus paper and/or one of many (120) references which he provided.

Late eighties marks the beginning of publications of semiconductor laser models based on drift-diffusion equations supplemented by Poisson equation used to describe electrons and holes plus relevant equations to describe photons, and also the interactions between optical field and carriers. The developed models were mostly 2-dimensional. Also the approach known as Device CAD (computer aided design) has been explicitly formulated. The approach is defined "as the technique of device design which solves general basic equations rigorously in order to accurately analyze devices under all kinds of operating conditions" [3].

Hitachi, Japan (1987)

That year marks publication [3] of a paper where HILADIES (HItachi LAser DIode Engineering Software) has been described along with first main results.It is a powerfull two-dimensional simulator which solves Poisson's equation, current continuity equations for electrons and holes as well as the wave equation and the rate equation for photons. Poisson's equation and the two continuity equations for electrons and holes were discretized using finite difference method (FDM). The wave equation was analyzed by the plane-wave expansion method. Heterojunctions and the carrier degeneracy were rigorously treated by introducing a band parameter and a Fermi-Dirac distribution model. The above equations were discretized by the finite difference method (FDM). Numerical solution was obtained using a supercomputer HITAC S-810. In the above mentioned paper, detailed steady-state numerical analysis have been performed for the channeled-substrate-planar (CSP) laser. Light output vs current characteristics including kinks and current vs voltage characteristics were determined. Reasonable agreement have been found between calculated and experimental results. Analysis of current leakage in InGaAsP/InP burried heterostructure lasers have been later reported by Hitachi group [4]. It was that no junction in the blocking layers is reverse-biased and that current confinement is due to electrically floating regions in the blocking structures.

Toshiba, Japan (1988)

In 1988 Toshiba has reported development of a two-dimensional device simulator to analyze optical and electrical properties of laser diodes [5].

The simulator can be used to model various kinds of laser diodes based on various materials and with arbitrary configurations. Optical, electrical, and thermal characteristics can be analyzed on a standard personal computer with a graphical user interface providing easy operation for general-purpose use. These device simulators are useful for the design and analysis of optical semiconductor devices such as laser diodes and light-emitting diodes.

The simulator has been extended to analyze optical, electrical and thermal characteristics of GaN based semiconductor lasers [6]. Recent summary of the work carried at Toshiba is reported in [7] and [8].

Eastman Kodak Company, USA (1988)

zmienic A fully self-consistent, steady-state, two-dimensional model of laser diode has been presented [9]. As in the above model, the electrical part consists of the simultaneous solution of the Poisson's and the electron and hole drift-diffusion equations for an arbitrary heterostructure device based on Lundstrom and Schuelke approach [10]. The model incorporates Fermi-Dirac statistics, position-dependent material and band parameters, and compositionally graded regions. The optical part consists of the solution of the field equation and the photon rate equation. Excellent agreement with experiment was obtained for both gain-guided and index-guided laser diodes. In that paper specific results were also given for channeled-substrate-planar (CSP) lasers. For example, it has been found that there can be significant lateral electron currents in the substrate region, that there can be a significant leakage of holes from the active layer, and that it is not necessary to grow on p-type substrates in order to avoid charging effects in the internal stripe of CSP lasers. For CSP lasers, it has also been determined that comparable electron current confinement can be provided by either zinc diffusion or an internal stripe; however, because zinc-diffused devices suffer less from the effects of hole current spreading, they are the most efficient devices overall.

UIUC, USA (1989)

At the University of Illinois at Urbana-Champain (UIUC) MINILASE have been under developed [11, 12] since the late eighties. Second generation of the simulator, Minilase-II [13] contains carrier dynamics including bulk transport, quantum carrier capture, spectral hole burning and quantum carrier heating. The issues of coupling of the electronic and optical parts [14], hot carrier effects [15] and tunnelling injection have been addressed. Comprehensive description of the results have been published recently [16]. In that publication, the authors report on the details of Minilase-II and the applications of the simulator to demonstrate the effects of various nonlinear processes occurring in quantum well lasers. Calculated modulation responses are compared directly with experimental data, showing good quantitative agreement.

NEC Corporation, Japan (1990)

In the paper [17] NEC Corporation group reported numerical studies on the self-aligned structure using two-dimensional simulator. Self-consistent calculations were carried out by coupling electrical and optical behavior. Drift-diffusion equations were solved in two-dimensional space by the finite difference method. Then, the 2D complex refractive permittivity in the active layer was determined which was incorporated into 2D wave equation. Wave equation was solved utilizing the finite element method and the stimulated-emission recombination distribution was determined. Finally, electrical behavior including output photon density distribution was computed.

The simulator was applied to AlGaAs-GaAs laser operating at $0.78 \mu m$ wavelength. Results were compared in detail with various experiments and good agreement was obtained. More recent work from NEC Corporation is reported in [18].

Ecole-Polytechnique, Montreal-Canada (1991).

They have developed the steady state model in two dimensions [19]. It is based on drift-diffusion semiconductor equations supplemented by the Poisson equation.

The authors have investigated the performance of double active region lasers. They studied the laser linearity and high characteristic temperature behavior as a function of the carrier concentration ratios in the two active layers. The devices were based on InGaAsP-InP and operated at $1.3 \mu m$ wavelength.

They were able to explain two of the most important particular features of such devices: their highly nonlinear L-I curves and high characteristic temperatures. They discovered the importance of the second active region which is related to the fact that the carriers may distribute themselves unevenly between the two active regions in the steady state.

Comprehensive description of the model is provided in [20]. The model has been applied in analyzing lateral carrier injection in multiple-quantum-well

DFB lasers [21]. The simulator has been further utilized in comparison between the Monte Carlo method and drift-diffusion approximation in quantum-well laser simulation [22]. Recent applications are discussed in [23].

NTT, Japan (1991)

About the same time, similar attempts were devoted at NTT [24–26]. They developed two-dimensional program based on solving electrical equations: Poisson's for electrostatic field and current continuity equations for electrons and holes. Because of large current densities in the active region (exceeding $10^{18}cm^{-3}$) those equations were solved using Fermi-Dirac statistics. Numerical solutions were obtained using the Gummel iteration scheme with a box discretization based on a five-point finite difference method. The optical characteristics were described using wave equation and the rate equation for phonons.

They analyzed the current blocking mechanism in InP buried heterostructure laser. The attention has been focused on the effects of stimulated emission on the current blocking capacity. They have also applied their simulator to DFB lasers. For a single electrode laser very good agreement in threshold characteristics has been found.

NRC, Canada (1992)

In 1992 first publication [27] of the sophisticated simulator developed at National Research Council (NRC), Ottawa, Canada has been published. The model includes self-consistent solution of electrical and optical equations, includes a wavelength- and position-dependent gain function and incorporates the effects of strain [28, 29]. It is based on drift-diffusion equations (for a summary see [30–32]). Several two-dimensional structures have been simulated using this model. Around 1994-95 the model was mature enough that it was capable of producing large amount of useful simulation data to help understand the operation of a complex laser device and to help optimize its design [32–34]. Over the time thermal effects have been added as well as longitudinal behaviour included [35]. Around 1995 main developer, Dr. Z.-M. (Simon) Li left NRC and formed company (currently CrossLight Software [36]). There the number of developments took place.

University of Toronto-Canada (1993)

Very sophisticated simulator, named FELES-1 (finite element light-emitter simulator) including thermal effects have been developed at University of Toronto [37]. It is fully self-consistent nonisothermal, two-dimensional which simultaneously solves Poisson's and drift-diffusion equations, along with the wave equation, photon rate equation and thermal conduction equation. The model includes the self-consistent solution of electrical, optical and thermal conduction equations using finite element method. In [37] an analysis has been presented for an AlGaAs-GaAs ridge laser diode and the results agreed well with available experimental data. A comparison made of the results between isothermal and nonisothermal simulations showed that the nonisothermal case has a higher threshold current and lower quantum efficiency than the idealized isothermal model. It has also been found that results depend critically on the thermal exchange boundary condition of the simulated device, demonstrating the importance of considering thermal exchange in the design of laser diodes. Extensive applications of this model has then been reported [38–40].

Bell Labs-USA (1994)

Year 1994 marks a publication of another powerfull laser simulator [41] and several applications [42]. The approach was to use program PADRE which solves semiconductor equations for heterostructures up to three spatial dimensions for an arbitrary doping distribution and arrangement of contacts. Soon, the simulator has been extended to quantum wells [43].

PADRE consists of a collection of robust programs for obtaining self-consistent solutions of the drift-diffusion equations. Other programs for calculating the optical intensity inside the laser, the capture and emission rates between bound and free carriers, the interaction between the confined carriers and the optical field, and a set of new, powerful schemes for obtaining rapid convergence of the non-linear equations have been added. In subsequent years, the model has been improved and extended [44]. Two key limitations of earlier laser simulations have been addressed: efficiency of the solution algorithm and the experimental verification of the simulation results. Extensive comparison of the simulation results to systematic experiments of InP based lasers have been made. Specifically, the role of p-doping profile have been studied both experimentally and theoretically [45] and [46]. In paper [47] the effect of p-doping on the temperature dependence of differential gain in FP and DFB lasers were analyzed. Simulations of carrier dynamics were reported in [48].

More recently, the simulator has been extended [49] to account for details of carrier transport, distribution of two-dimensional carriers within the quantum well, optical gain spectra and photon rate equations. Those results were reported in more recent papers [50, 51] and [52].

University of Wales-U.K. (1994)

They presented a fully two-dimensional self-consistent numerical model [53] of the steady-state behavior of constricted-mesa InGaAsP/InP buried-heterostructure lasers. The model enables investigation of temperature characteristics. The temperature dependence was included in the Fermi-Dirac statistics, bandgaps, mobilities, densities of states, Auger recombination coefficients, intervalence band absorption, optical gain and thermal conductivities. It was used for optimum design of reduced threshold current and other characteristics. More results were published later [54] where the authors demonstrated the influence of the whole device structure on the characteristics of GaInP lasers.

Technikal University of Lodz, Poland (1994-1995)

A new comprehensive thermal-electrical self-consistent model of proton-implanted top-surface-emitting lasers was described [55]. Next year, they reported on the extension of the above model to account for carrier diffusion [56]. That comprehensive, fully self-consistent thermal-electrical finite-element model was used to investigate thermal properties of proton-implanted top-surface-emitting lasers. Temperature dependence of material and device parameters was taken into account, and multiple heat sources were considered. The analysis demonstrated that carrier diffusion influences strongly the distribution of the main heat source located in the active region. Further progress was reported in a series of papers [57–65]. Paper [66] summarizes progress of that group. The self-consistent optical-electrical-thermal-gain threshold model of the oxide-confined (OC) quantum-dot (QD) (InGa)As-GaAs vertical-cavity surface-emitting diode laser (VCSEL) is demonstrated. The model has been developed to enable better understanding of physics of an operation of GaAs-based OC QD VCSELs in a full complexity of many interactions in its volume between individual physical phenomena. In addition, the model has been applied to design and optimize the low-threshold long-wavelength 1.3-/spl mu/m GaAs-based OC QD VCSELs for the second-generation optical-fiber communication systems and to examine their anticipated room-temperature (RT) performance. An influence of many construction parameters on device RT lasing thresholds and mode selectivity has been investigated. Some essential design guidelines have been proposed to support efforts of technological centers in producing low-threshold single-mode RT devices.

University of New Mexico (1994-1995)

The University of New Mexico group collaborated closely with the Technikal University of Lodz [55, 56]. The original approach taken was not literally based on the drift-diffusion model but it is an important development for VCSELs since thermal effects which are important in those systems were included. In a later paper [67] report on a versatile electro-thermal and optical numerical simulation tool has been provided. Further results obtained in collaboration with CFD Research Corporation were reported [68]. A self-consistent electro-thermo-opto three-dimensional model had been developed. Temperature dependency of critical device and material properties were included, as well as multiple heat generation mechanisms. Results of extensive simulations were reported over period of 2003-05. Those included analysis of InGaN-based ultraviolet emitting heterostructures with quaternary AlInGaN barriers [69], high-performance InGaAs-GaAs-AlGaAs broad-area diode lasers with impurity-free intermixed active region [70], self-consistent calculation of current self-distribution effect in GaAs-AlGaAs oxide-confined VCSELs [71] and effects of resonant mode coupling on optical characteristics of InGaN-GaN-AlGaN lasers [72].

University of Waterloo-Canada (1995)

In that year a paper on a two-dimensional numerical model for DFB semiconductor laser simulator made of bulk material has been published [73]. Their model includes the transverse carrier transport and longitudinal spatial hole-burning effects in a bulk DFB lasers. The model solves self-consistently the carrier transport equations in a hetero-diode in the transversal dimension and the optical wave equation in the lateral dimension, together with the phonon rate equation. Transversal field distribution was calculated only once for a reference waveguide. The transversal modal gain, effective-index change and modal loss were considered using the first-order perturbation theory.

Later-on the model has been extended [74] and a comprehensive physics-based 3D model of DFB laser diode has been reported. New model included optical field confinement, the carrier transport and the heat transfer over the cross section of the laser. It also accounted for the longitudinal spatial hole-burning effect.

ETH – Swiss Federal Institute of Technology, Zurich-Switzerland (1996)

In 1996 device simulator for smart integrated systems (DESSIS) [75] was described. In that paper the most important achievements of the ESPRIT 6075 (DESSIS) project were presented. The project was carried out in the period 1992-1995 with the participation of SGS-Thomson Microelectronics, Robert Bosch GmbH, the Swiss Federal Institute of Technology and the University of Bologna. Within the framework of this project, a new mixed-mode circuit/device simulator called DESSIS has been developed.

The extension of DESIIS to a multidimensional semiconductor laser simulator has been reported few year later [76, 77]. It follows a rate equation approach for the coupling between optics and electronics. The electronic and optical equations were solved in a self-consistent way for one, two, and three dimensions. The ETH is one of the leading groups in the field of advanced simulations of semiconductor lasers. More details in PhD Thesis [78–80]. Dynamic characteristics under large-signal modulation are reported in [81].

Wilfrid Laurier University-Canada (1996)

The drift-diffusion approach was used to describe carrier transport effects [82] of MQW semiconductor lasers. Standard drift-diffusion equations for electrons and holes were combined with one-dimensional wave equation and photon rate equation. Those equations were solved self-consistently in the steady-state and compared with rate equations approach. Later-on, we have extended that approach to simulate distributed feedback (DFB) semiconductor lasers. Particular attention has been paid to analyze large-signal response and the wavelength chirp [83] in multiple quantum well semiconductor lasers.

SimWindows (1998)

SimWindows is an easy to use, 1D device simulator written by David Winston at the Optoelectronics Computer Systems Center of the University of Colorado [84, 85]. You can download it from the following web site [86]. SimWindows allows to simulate operation of optoelectronic semiconductor devices (not only semiconductor lasers) where electrical, optical, and thermal effects are coupled.

The simulator combines the major physical models necessary for simulating optoelectronic devices such as surface emitting lasers and quantum well solar cells. The software extends many of the traditional electrical models by adding effects such as quantum confinement, tunneling current, and complete Fermi-Dirac statistics. The optical model includes computing electromagnetic field reflections at interfaces and determining the resonant frequency of laser cavities. SimWindows implements two thermal models which either compute the lattice temperature or the electron temperature. This combination of models can predict many aspects of optoelectronic devices that traditional simulation programs can not. SimWindows provides a large degree of flexibility for the user to control and modify the models in SimWindows. The user can also add new models depending on the circumstances.

Tsinghua University, Stanford University (1998)

The group had implemented a practical two-dimensional simulator of quantum-well lasers [87]. New, very stable and fast method has been developed. The electric equations, scalar Helmholtz equation and photon rate equation were solved self-consistently. In addition, the carrier energy transport and lattice thermal diffusion are accounted for. The method has been applied to the simulations of a GRIN-SCH BH SQW laser. Good agreement with experimental data was found.

Weierstrass Institute for Applied Analysis and Stochastics-WIAS (2000)

WIAS - Weierstrass Institute for Applied Analysis and Stochastics, Berlin-Germany maintains strong and active group devoted to simulations of nano- and optoelectronic components.

In the paper [88] the results of simulations of edge emitting RW lasers based on InP were reported. The simulator called TeSCA employs two-dimensional solver of carrier transport equations within drift-diffusion approach in combination with the Poisson equation and Fermi-Dirac statistics. The heating effects were considered within a lattice temperature based model.

Further results are reported in [89].

WIAS-TeSCA [91] (Two- and three-dimensional semiconductor analysis package) is a program system for the numerical simulation of charge transfer processes in semiconductor structures, especially also in semiconductor lasers. It is based on the drift-diffusion model and considers a multitude of additional physical effects, like optical radiation, temperature influences and the kinetics of deep (trapped) impurities. Its efficiency is based on the analytic study of the strongly nonlinear system of partial differential equations (van Roosbroeck), which describes the electron and hole currents. Very efficient numerical procedures for both the stationary and transient simulation have been implemented in WIAS-TeSCA.

B.1 Companies

Several companies develop and sell simulators of active photonic devices and semiconductor lasers in particular. Here, we provide short summary of those activities.

Crosslight Software

Crosslight Software Inc. [36] (formerly Beamtek Software) is an international company with headquarters in Vancouver, Canada. The Company provides simulation software for modelling semiconductor devices and processes, including semiconductor lasers and other optoelectronic devices. The software can model many practical devices, like Fabry-Perot lasers and VCSELs with thermal models.

The Company has a broad line of software products. For the development of PICS3D, a Photonic Integrated Circuit Simulator in 3D, they had received Commercial Technology Achievement Award from Laser Focus World in 1998.

CFDRC

CFD Research Corporation [90] which on August 7, 2007 celebrated its 20th anniversary, provides integrated software tools and modelling expertise for multiscale, multiphysics design of variety of optoelectronic devices. Extensive information is provided at the Company home page [90] and includes quantum-well (QW) based light sources (VCSELs, Edge-Emitting Lasers), photodetectors (MSM, p-i-n), and others. The multiphysics modeling of intracavity-contacted VCSEL is reported by Riely et al [92]. The group developed a comprehensive multiphysics modeling tool known as CFD-ACE+ O'SEMI. The modeling tool integrates electronic, optical, thermal and material gain data models for the design of VCSELs and edge emitting lasers.

Another tool, the CFDRC NanoTCAD Device Simulator is based on hydrodynamic self consistent model of carriers and energy transport. CFDRC NanoTCAD device simulator was recently extended by adding the quantum-corrected DD model [93], which is fast and multidimensional (2D and 3D). This model has successfully reproduced negative differential resistance (NDR) for Esaki tunnel junction diode. Also, kinetic model with quantum-corrected potential have been successfully developed and tested at CFDRC for modelling of non-equilibrium and quantum carrier transport in semiconductor nano-devices [93]. This model involves a solution of 4D Boltzmann transport equation (BTE).

RSoft

Rsoft Design Group offers LaserMOD [94] which is a photonic device design software tool for simulating the optical, electronic, and thermal properties of semiconductor lasers. The software comes with material libraries and tutorial examples. Device applications currently include: edge emitting lasers, such as Fabry-Perot and DFB type VCSEL and silicon modulators. More information about this and also other products can be found on the Company web page.

Silvaco

Silvaco company [95] has been developing semiconductor laser diode simulator for some time. On their web page [96] they say: " Laser is the world's first commercially available simulator for semiconductor laser diodes. Laser works in conjunction with Blaze in the ATLAS framework to provide numerical solutions for the electrical behavior (DC and transient responses) and optical behavior of edge emitting Fabry-Perot type lasers diodes."

Another product, as informed on Silvaco web page, is the Vertical Cavity Surface Emitting Laser Simulation (VCSEL) [97]. It is is used in conjunction with the ATLAS framework to produce physically based simulations of vertical cavity surface emitting lasers. VCSEL joins sophisticated device simulation to obtain electrical and thermal behavior with state of the art models for optical behavior.

NextNano

NextNano [98] was founded in August 2004 as a spin-off from the Walter Schottky Institute of the Technical University of Munich, Germany. The Company develops software for the simulation of electronic and optoelectronic semiconductor nano devices and materials.

As explain in their Executive Summary, the Company proposes better physical methods for the calculation of the quantum mechanical properties of an arbitrary combination of geometries and materials.

PhotonDesign

Photon Design [99] exists since 1992. Their goal was to to provide professional quality software to the photonics industry. Currently, they provide a wide range of photonics tools for both active and passive photonics components. They advertise several products including CLADISS which is a longitudinal laser diode model. CLADISS was developed by the University of Gent over many years and has been the source of many publications. It holds a high reputation among the international modelling community. The model of CLADISS was described in [100] and used for the analysis of multisection diode lasers. The simulator can carry out a threshold, DC, AC, and a noise analysis.

Semiconductor Technology Research (STR)

Semiconductor Technology Research, Inc. (STR) [101] provides dedicated software and consulting services for simulation and optimization of crystal growth techniques and for modelling of semiconductor based devices. They expertiese include modeling of advanced semiconductor devices, such as light emitting diodes, laser diodes, heterojunction bipolar transistors, high-electron mobility transistors and Schottky diodes. One of their products is the Simulator of Light Emitters based on Nitride Semiconductors (SiLENSe). It offers the ability of modelling band diagrams and characteristics of light emitting diodes (LEDs). The LED operation is considered within the 1D drift-diffusion model of carrier transport in the heterostructure that accounts for specific features of the nitride semiconductors [102].

Synopsys

Synopsys [103] has over 20 years of existence. Over those years, they provided solutions to difficult problems facing engineers who are pushing electronic design to the limit. The company provides professional services as well as sells many products.

Sentaurus Device Optoelectronics is an advanced device simulator capable of simulating a wide range of semiconductor devices. It includes state-of-the-art numeric solvers and a comprehensive set of models for carrier and heat transport, quantization effects, and heterostructures.

The Opto option supports the simulation of light-emitting devices with advanced band structure and gain calculations. The EMW option allows full-wave solutions of the Maxwell equations to account for physical optics in advanced devices.

Integrated Systems Engineering AG

In 1993 ISE (Integrated Systems Engineering AG) [104] was founded as a spin-off company by the members of the Integrated Systems Laboratory (IIS)of the Swiss Federal Institute of Technology Zurich (ETH Zurich). It is now part of Synopsys [103].

One of the important products developed at ISE AG was DESSIS. DESSIS (Device Simulation for Smart Integrated Systems) is the device simulator of ISE AG which was available in version 7.5 (as of 2003). DESSIS is a multidimensional simulator for simulation of one, two and three-dimensional semiconductor devices and has mixed-mode capabilities for circuit simulation. Depending on the device which is simulated and on the accuracy which is needed different transport models can be used, which are drift-diffusion, self-heating, hydrodynamic, or Monte-Carlo which is performed in a user-specified window only. DESSIS is capable to perform steady state, transient, AC-small signal, and noise analysis. Mixed mode simulations can be carried out by combining physical devices and compact models in a circuit.

Summary of that simulator is provided in [105]

B.2 More Recent Developments

Nowadays, there are intensive activities around the globe on active semiconductor devices. Those are regularly reported. Also, there are periodic conferences related to simulations of active optoelectronic devices. They will be now briefly summarized.

Numerical Simulation of Optoelectronic Devices (NUSOD)

List of current up-to-date software directory related to photonic and optoelectronic devices is provided at the web site of the NUSOD Institute [106]. The general mandate of the NUSOD Institute as explained at their web page is to connect theory and practice in optoelectronics. For that purpose numerical simulation is an excellent tool. However, one should be very careful on that path, since in order to achieve realistic results models used along with the input parameters should be very carefully evaluated.

Starting with year 2000 NUSOD Institute organizes yearly conferences on Numerical Simulations of Semiconductor Optoelectronic Devices (NUSOD) (first organized by Dr. J. Piprek while he was at Electrical and Computer Engineering Department, Santa Barbara). Some of the materials from those meetings were published [107, 108].

International Workshop on Computational Electronics (IWCE)

International Workshop on Computational Electronics [109] covers all aspects of advanced simulations in electronics and optoelectronics. Typical topics discussed involved: simulations of optical processes and optoelectronic devices, quantum transport and dynamics of nanostructures, molecular and organic electronics. The scientific program, organized in a single-session format, consists of invited lectures, contributed talks, and poster presentations.

The workshop is intended to be an international forum for discussions on the current trends and future directions of computational electronics. The emphasis of the contributions is on interdisciplinary aspects of Computational Electronics, touching Physics, Engineering, Applied Mathematics, as well as Chemistry and Biology.

Society of Photo-Optical Instrumentation Engineers (SPIE) Proceedings

For several years there had been published papers as Proceedings of SPIE presented at various conferences under general title "Physics and Simulation of Optoelectronic Devices". They provided extensive and state-of-the-art results of new developments and also applications of commercial software to the design and analysis of new structures. The following have been published over the years 1992 to 2007, first edited by D. Yevick and then by M. Osinski and collaborators: [110–139]

Special Issues of IEEE Journal on Selected Topics of Quantum Electronics

From time to time there are special issues of IEEE Journal on Selected Topics of Quantum Electronics devoted to issues on semiconductor lasers which document current state of semiconductor lasers including papers on design and modelling of optoelectronic devices. Several recent examples are: [140–144].

Bibliography

[1] J. Hecht. The breakthrough birth of the diode laser. *Optics and Photonics News*, 18:38–43, 2007.

[2] J. Buus. Principles of semiconductor laser modelling. *Proc. IEE Pt.J*, 132:42 – 51, 1985.

[3] T. Ohtoshi, K. Yamaguchi, C. Nagaoka, T. Uda, Y. Murayama, and N. Chinone. A two-dimensional device simulator of semiconductor lasers. *Solid State Electron.*, 30:627–638, 1987.

[4] T. Ohtoshi, K. Yamaguchi, and N. Chinone. Analysis of current leakage in InGaAsP/InP buried heterostructure lasers. *IEEE J. Quantum Electron.*, 25:1369–1375, 1989.

[5] G. Hatakoshi, M. Kurata, E. Iwasawa, and N. Motegi. General two-dimensional sevice simulator for laser diodes. *Transactions of the IEICE Japan*, E71:923–925, 1988.

[6] G. Hatakoshi, M. Onomura, and M. Ishikawa. Optical, electrical and thermal analysis for GaN semiconductor lasers. *Int. J. Numer. Model.*, 14:303–323, 2001.

[7] *Optical, electrical and thermal simulation for semiconductor lasers and light-emitting diodes*, Proceedings of the IEEE/LEOS 3rd International Conference on Numerical Simulation of Semiconductor Optoelectronic Devices, 2003.

[8] G. Hatakoshi, Y. Hattori, S. Saito, N. Shida, and S. Nunoue. Device simulator for designing high-efficiency light-emitting diodes. *Jpn. J. Appl. Phys.*, 46:5419–5425, 2007.

[9] Keith B. Kahen. Two-dimensional simulation of laser diodes in the steady state. *IEEE J. Quantum Electron.*, 24:641–651, 1988.

[10] M.S. Lundstrom and R.J. Schuelke. Numerical analysis of heterostructure semiconductor devices. *IEEE Trans. Electron Devices*, ED-30:1151–1159, 1983.

[11] G.H. Song, K. Hess, T. Kerkhoven, and U. Ravaioli. Two-dimensional simulator for semiconductor lasers. *International Electron Devices Meeting,3-6 Dec. 1989*, pages 143 – 146, 1989.

[12] G.H. Song. *Two-dimensional simulation of quantum-well lasers including energy transport.* PhD thesis, University of Illinois at Urbana-Champaign, 1990.

[13] Grupen M. E. *The self-consistent simulation of carrier transport and its effect on the modulation response in semiconductor quantum well lasers.* PhD thesis, University of Illinois at Urbana-Champaign, 1994.

[14] M. Grupen, U. Ravaioli, A. Galick, K. Hess, and T. Kerkhoven. Coupling the electronic and optical problems in semiconductor quantum well laser simulations. *Proc. SPIE*, 2146:133, 1994.

[15] M. Grupen and K. Hess. Hot carrier effects in conventional injection and tunneling injection quantum well laser diodes. *Proc. SPIE*, 2994:474, 1997.

[16] M. Grupen and K. Hess. Simulation of carrier transport and nonlinearities in quantum-well laser diodes. *IEEE J. Quantum Electron.*, 34:120–140, 1998.

[17] M. Ueno, S. Asada, and S. Kumashiro. Two-dimensional numerical analysis of lasing characteristics for self-aligned structure semiconductor lasers. *IEEE J. Quantum Electron.*, 26:972–981, 1990.

[18] M. Ueno, S. Asada, and S. Kumashiro. Two-dimensional analysis of astigmatism in self-aligned structure semiconductor laser. *IEEE J. Quantum Electron.*, 28:1487–1495, 1992.

[19] A. Champagne, R. Maciejko, and J.M. Glinski. The performance of double active region InGaAsP lasers. *IEEE J. Quantum Electron.*, 27:2238–2247, 1991.

[20] A. Champagne. *Modelisation des lasers InGaAsP-InP a double heterostructure et a double region active*. PhD thesis, Departement de Genie Physique, Ecole Polytechnique, Universite de Montreal, 1992.

[21] A. Champagne, R. Maciejko, and T. Makino. Enhanced carrier injection efficiency from lateral current injection in multiple-quantum-well DFB lasers. *IEEE J. Quantum Electron.*, 8:749–751, 1996.

[22] A.D. Guclu, R. Maciejko, A. Champagne, and M. Abou-Khalil. Comparison between the Monte Carlo method and drift-diffusion approximation in quantum-well laser simulation. *J. Appl. Phys.*, 84:4673–4676, 1998.

[23] A. Champagne, J. Camel, R. Maciejko, K.J. Kasunic, D.M. Adams, and B. Tromborg. Linewidth broadening in a distributed feedback laser integrated with a semiconductor optical amplifier. *IEEE J. Quantum Electron.*, 38:1493–1502, 2002.

[24] T. Yamanaka, S. Seki, and K. Yokoyama. Numerical analysis of static wavelength shift for DFB lasers with longitudinal mode spatial hole burning. *IEEE Photon. Technol. Lett.*, 3:610–612, 1991.

[25] S. Seki, T. Yamanaka, and K. Yokoyama. Two-dimensional numerical analysis of current blocking mechanism in InP buried heterostructure lasers. *J. Appl. Phys.*, 71:3572–3578, 1992.

[26] K. Yokoyama, T. Yamanaka, and S. Seki. Two-dimensional numerical simulator for multielectrode distributed feedback laser diodes. *IEEE J. Quantum Electron.*, 29:856–863, 1993.

[27] Z.-M. Li, K.M. Dzurko, A. Delage, and S.P. McAlister. A self-consistent two-dimenional model of quantum-well semiconductor lasers: optimization of a GRIN-SCH SQW laser structure. *IEEE J. Quantum Electron.*, 28:792–803, 1992.

[28] Z-M. Li, M. Dion, S.P. McAlister, R.L. Williams, and G.C. Aers. Incorporating of strain into a two-dimensional model of quantum-well semiconductor lasers. *IEEE J. Quantum Electron.*, 29:346–354, 1993.

[29] Li Z.-M., M. Dion, Y. Zou, J. Wang, M. Davies, and S.P. McAlister. An approximate k·p theory for optical gain of strained InGaAsP quantum-well lasers. *IEEE J. Quantum Electron.*, 30:538–546, 1994.

[30] S.P. McAlister and Z.-M. Li. Two dimensional simulation of quantum-well lasers. In W.W. Chow and M. Osinski, editors, *Physics and Simulation of Optoelectronic Devices II*, volume 2146 of *Proc. SPIE*, pages 162–173, 1994.

[31] M. Dion, Z-M. Li, D. Ross, F. Chatenoud, R.L. Williams, and S. Dick. A study of the temperature sensitivity of GaAs-(Al,Ga)As multiple quantum-well GRINSCH lasers. *IEEE J. Select. Topics Quantum Electron.*, 1:230–233, 1995.

[32] Z.-M. Li. Two-dimensional numerical simulation of semiconductor lasers. In W.P. Huang, editor, *Electromagnetic Waves, Methods for Modeling and Simulation of Guided-Wave Optoelectronic Devices: Part II:Waves and Interactions*, volume PIER 11, pages 301–344. 1995.

[33] Z.-M. Li and T. Bradford. A comparative study of temperature sensitivity of InGaAsP and AlGaAs MQW lasers using numerical simulations. *IEEE J. Quantum Electron.*, 31:1841–1847, 1995.

[34] K.-W. Chai, Z-M. Li, S.P. McAlister, and J.G. Simmons. Numerical drift-diffusion simulation of Auger hot electron transport in InGaAsP/InP double heterostructure laser diodes. *International Journal of Numerical Modelling: Electrical Networks, Devices and Fields*, 7:267–281, 1994.

[35] Z.-M. (Simon) Li. Physical models and numerical simulation of modern semiconductor lasers. In M. Osinski and W.W. Chow, editors, *Physics and Simulation of Optoelectronic Devices V*, volume 2994 of *Proc. SPIE*, pages 698–708, 1997.

[36] For actual update, visit company at www.crosslight.com.

[37] G.-L. Tan, N. Bewtra, K. Lee, and J.M. Xu. A two-dimensional non-isothermal finite element simulation of laser diodes. *IEEE J. Quantum Electron.*, 29:822–835, 1993.

[38] N. Bewtra, D.A. Suda, G.-L. Tan, F. Chatenoud, and J.M. Xu. Modeling of quantum-well lasers with electro-opto-thermal interaction. *IEEE J. Select. Topics Quantum Electron.*, 1:331–340, 1995.

[39] E.H. Sargent, G.-L. Tan, and J.M. Xu. Physical model of OEIC-compatible lateral current injection lasers. *IEEE J. Select. Topics Quantum Electron.*, 3:507–512, 1997.

[40] G.-L. Tan, R.S. Mand, and J.M. Xu. Self-consistent modeling of beam instabilities in 980-nm fiber pump lasers. *IEEE J. Quantum Electron.*, 33:1384–1395, 1997.

[41] R.F. Kazarinov and M.R. Pinto. Carrier transport in laser heterostructures. *IEEE J. Quantum Electron.*, 30:49–53, 1994.

[42] R.F. Kazarinov and G.L. Belenky. Novel design of AlGaInAs-InP lasers operating at $1.3\mu m$. *IEEE J. Quantum Electron.*, 31:423–426, 1995.

[43] M.A. Alam, M.S. Hybertsen, R.K. Smith, G.A. Baraff, and M.R. Pinto. Simulation of semiconductor quantum well lasers. In *Proc. SPIE*, volume 2994, pages 709–722, 1997.

[44] M.S. Hybertsen, M.A. Alam, G.A. Baraff, A.A. Grinberg, and R.K. Smith. Role of non-equilibrium carrier distributions in multi-quantum well InGaAsP based lasers. In *Proc. SPIE*, volume 3283, pages 379–383, 1998.

[45] G.L. Belenky, C.L. Reynolds, D.V. Donetsky, G.E. Shtengel, M.S. Hybertsen, M.A. Alam, G.A. Baraff, R.K. Smith, R.F. Kazarinov, J. Winn, and L.E. Smith. Role of p-doping profile and regrowth on the static characteristics of $1.3\mu m$ MQW InGaAsP-InP lasers: experiment and modeling. *IEEE J. Quantum Electron.*, 35:1515–1520, 1999.

[46] M.S. Hybertsen, M.A. Alam, G.E. Shtengel, G.L. Belenky, C.L. Reynolds, D.V. Donetsky, R.K. Smith, G.A. Baraff, R.F. Kazarinov, J.D. Wynn, and L.E. Smith. Simulation of semiconductor quantum well lasers. In *Proc. SPIE*, volume 3625, pages 524–534, 1999.

[47] G.L. Belenky, C.L. Reynolds, L. Shterengas, M.S. Hybertsen, D.V. Donetsky, G.E. Shtengel, and S. Luryi. Semiconductor laser diodes. *IEEE Photon. Technol. Lett.*, 12:969–971, 2000.

[48] M.S. Hybertsen, M.A. Alam, G.A. Baraff, R.K. Smith, G.E. Shtengel, C.L. Reynolds, and G.L. Belenky. Simulation of carrier dynamics in multi-quantum well lasers. In *Proc. SPIE*, volume 3944, pages 486–491, 2000.

[49] M.A. Alam, M.S. Hybertsen, R.K. Smith, and G.A. Baraff. Simulation of semiconductor quantum well lasers. *IEEE Trans. Electron Devices*, 47:1917–1925, 2000.

[50] M.S. Hybertsen, B. Witzigmann, M.A. Alam, and R.K Smith. Role of carrier capture in microscopic simulation of multi-quantum-well semiconductor laser diodes. *J. Comp. Electron.*, 1:113–118, 2002.

[51] B. Witzigmann, M.S. Hybertsen, C.L. Lewis Reynolds, Jr., G.L. Belenky, L. Shterengas, and G.E. Shtengel. Microscopic simulation of the temperature dependence of static and dynamic $1.3\mu m$ multi-quantum-well laser performance. *IEEE J. Quantum Electron.*, 39:120–129, 2003.

[52] B. Witzigmann and M.S. Hybertsen. A theoretical investigation of the characteristic temperature t_0 for semiconductor lasers. *IEEE J. Select. Topics Quantum Electron.*, 9:807–815, 2003.

[53] M. Gault, P. Mawby, A.R. Adams, and M. Towers. Two-dimensional simulation of constricted-mesa InGaAsP/InP burried-heterostructure lasers. *IEEE J. Quantum Electron.*, 30:1691–1700, 1994.

[54] P. Blood, D.L. Foulger, P.M. Smowton, and P. Mawby. Simulation of GaInP laser structures. In M. Osinski and W.W. Chow, editors, *Physics and Simulation of Optoelectronic Devices V*, volume 2994 of *Proc. SPIE*, pages 736–746, 1997.

[55] W. Nakwaski and M. Osinski. Self-consistent thermal-electrical modeling of proton-implanted top-surface-emitting semiconductor lasers. In W.W. Chow and M. Osinski, editors, *Physics and Simulation of Optoelectronic Devices II*, volume ..... of *Proc. SPIE*, pages 34–43, 1994.

[56] R.P. Sarzala, W. Nakwaski, and M. Osinski. Finite-element comprehensive thermal modeling of proton-implanted top-surface-emitting lasers. *Lasers and Electro-Optics Society Annual Meeting, 1995. 8th Annual Meeting Conference Proceedings*, 2:437 – 438, 1995.

[57] W. Nakwaski and R.P Sarzala. Analysis of transverse modes in gain-guided vertical-cavity surface-emitting lasers. *Lasers and Electro-Optics Society Annual Meeting, 1997. IEEE LEOS '97 10th Annual Meeting. Conference Proceedings.*, 2:293 – 294, 1997.

[58] R.P. Sarzala and W. Nakwaski. Carrier diffusion inside active regions of gain-guided vertical-cavity surface-emitting lasers. *IEE Proceedings-Optoelectronics*, 144:421 – 425, 1997.

[59] M. Wasiak, R.P. Sarzala, T. Czyszanowski, P. Mackowiak, W. Nakwaski, and M. Bugajski. InGaAs/GaAs quantum-dot diode lasers for $1.3\mu m$ optical fibre communication. *Proceedings of the 2002 4th International Conference on Transparent Optical Networks, 2002.*, 1:144 – 147, 2002.

[60] R.P. Sarzala, P. Mendla, M. Wasiak, P. Mackowiak, W. Nakwaski, and M. Bugajski. Three-dimensional comprehensive self-consistent simulation of a room-temperature continuous-wave operation of GaAs-based

1.3μm quantum-dot (InGa)As/GaAs vertical-cavity surface-emitting lasers. *Transparent Optical Networks, 2003. Proceedings of 2003 5th International Conference on*, 2:95 – 98, 2003.

[61] R.P. Sarzala, P. Mackowiak, M. Wasiak, T. Czyszanowski, and W. Nakwaski. Structure optimisation of 1.3μm (GaIn)(NAs)/GaAs in-plane lasers. *IEE Proceedings-Optoelectronics*, 150:56 – 58, 2003.

[62] R.P. Sarzala, P. Mackowiak, M. Wasiak, T. Czyszanowski, and W. Nakwaski. Simulation of performance characteristics of GaInNAs vertical-cavity surface-emitting lasers. *IEE Proceedings-Optoelectronics*, 150:83 – 85, 2003.

[63] W. Sarzala, R.P.; Nakwaski. Optimisation of gaas-based (GaIn)(NAs)/GaAs vertical-cavity surface-emitting diode lasers for high-temperature operation in 1.3μm optical-fibre communication systems. *IEE Proceedings-Optoelectronics*, 151:417 – 420, 2004.

[64] W. Nakwaski and R.P. Sarzala. Mode selectivity in oxide-confined vertical-cavity surface-emitting lasers. *Proceedings of the 5th International Conference on Numerical Simulation of Optoelectronic Devices, 2005. NUSOD '05.*, ...:17 – 18, 2005.

[65] R.P. Sarzala and W. Nakwaski. Comparative analysis of various designs of oxide-confined vertical-cavity surface-emitting diode lasers. *2006 International Conference on Transparent Optical Networks*, 4:160 – 163, 2006.

[66] R.P. Sarzala. Modeling of the threshold operation of 1.3μm GaAs-based oxide-confinement (InGa)As-GaAs quantum-dot vertical-cavity surface-emitting lasers. *IEEE J. Quantum Electron.*, 40:629–639, 2004.

[67] V.A. Smagley, G.A. Smolyakov, P.G. Eliseev, M. Osinski, and A.J. Przekwas. Current self-distribution effect in vertical-cavity surface-emitting semiconductor lasers. In M. Osinski, P. Blood, and A. Ishibashi, editors, *Physics and Simulation of Optoelectronic Devices VI*, volume 3283 of *Proc. SPIE*, pages 171–182, 1998.

[68] Yu, E. and Osinski, M. and Nakwaski, W. and Turowski, M. and Przekwas, A.J. Thermal crosstalk in arrays of proton-implanted top-surface emitting lasers. In M. Osinski, P. Blood, and A. Ishibashi, editors, *Physics and Simulation of Optoelectronic Devices VI*, volume 3283 of *Proc. SPIE*, pages 384–395, 1998.

[69] J. Lee, P.G. Eliseev, M. Osinski, D.-S. Lee, D.I. Florescu, S. Guo, and M. Pophristic. Ingan-based ultraviolet emitting heterostructures with quaternary AlInGaN barriers. *IEEE J. Select. Topics Quantum Electron.*, 9:1239–1245, 2003.

[70] Y.-R. Zhao, G.A. Smolyakov, and M. Osinski. High-performance ingaas-gaas-algaas broad-area diode lasers with impurity-free intermixed active region. *IEEE J. Select. Topics Quantum Electron.*, 9:1333–1339, 2003.

[71] M. Osinski, V.A. Smagley, M. Lu, G.A. Smolyakov, P.G. Eliseev, B.P. Riely, P.H. Shen, and G.J. Simonis. Self-consistent calculation of current self-distribution effect in gaas-algaas oxide-confined vcsels. *IEEE J. Select. Topics Quantum Electron.*, 9:1422–1430, 2003.

[72] G.A. Smolyakov, P.G. Eliseev, and M. Osinski. Effects of resonant mode coupling on optical characteristics of ingan-gan-algan lasers. *IEEE J. Quantum Electron.*, 41:517–524, 2005.

[73] A.D. Sadovnikov, X. Li, and W.-P. Huang. A two-dimensional DFB laser model accounting for carrier transport effects. *IEEE J. Quantum Electron.*, 31:1856–1862, 1995.

[74] X. Li, A.D. Sadovnikov, W.-P. Huang, and T. Makino. A physics-based three-dimensional model for distributed feedback laser diodes. *IEEE J. Quantum Electron.*, 34:1545–1553, 1998.

[75] G. Baccarani, M. Rudan, M. Lorenzini, W. Fichtner, J. Litsios, A. Schenk, P. van Staa, L. Kaeser, A. Kampmann, A. Marmiroli, C. Sala, and E. Ravanelli. Device simulation for smart integrated systems dessis. In *Proceedings of the Third IEEE International Conference on Electronics, Circuits, and Systems,ICECS '96.*, volume 2, pages 752 – 755, 1996.

[76] B. Witzigmann, A. Witzig, and W. Fichtner. A full 3-dimensional quantum well laser simulator. In *7th International Workshop on Computational Electronics, 2000. Book of Abstracts. IWCE Glasgow 2000.22-25 May 2000*, pages 13–14, 2000.

[77] B. Witzigmann, A. Witzig, and W. Fichtner. A multidimensional laser simulator for edge-emitters including quantum carrier capture. *IEEE Trans. Electron Devices*, 47:1926–1934, 2000.

[78] Wolfgang Fichtner, Qiuting Huang, Heinz Ji.ckel, Gerhard Tr.ster, and Bernd Witzigmann, editors. *Industrial-Strength Simulation of Quantum-Well Semiconductor Lasers*. Hartung-Gorre Verlag Inh. Dr. Renate Gorre D-78465 Konstanz Germany, 2005.

[79] Wolfgang Fichtner, Qiuting Huang, Heinz Jœckel, Gerhard Trïster, and Bernd Witzigmann, editors. *Physics and Simulation of Semiconductro Lasers: Static and Dynamic Chracteristics*. Hartung-Gorre Verlag Inh. Dr. Renate Gorre D-78465 Konstanz Germany, 2006.

[80] Wolfgang Fichtner, Qiuting Huang, Heinz Jäckel, Gerhard Tröster, and Bernd Witzigmann, editors. *Multidimensional modeling and simulation of wavelength-tunable semiconductor lasers*. Hartung-Gorre Verlag Inh. Dr. Renate Gorre D-78465 Konstanz Germany, 2006.

[81] S. Odermattnd, B. Witzigmann, and B. Schmithuesen. Harmonic balance analysis for semiconductor lasers under large-signal modulation. *Optical and Quantum Electronics*, 38:1039–1044, 2006.

[82] Y. Chen, M.S. Wartak, H. Lu, and T. Makino. Rate equation description of carrier transport effects in multiple quantum well lasers. In P.C. Chen and T.D. Milster, editors, *Laser Diode Chip and Packaging Technology*, volume 2610 of *Proc. SPIE*, pages 34–43, 1996.

[83] Y. Chen and M.S. Wartak. Wavelength chirp of DFB lasers with carrier transport. *Microwave and Optical Technology Letters*, 15:291–294, 1997.

[84] D.W. Winston. *Physical Simulation of Optoelectronic Semiconductor Devices*. PhD thesis, University of Colorado at Boulder, 1996.

[85] D.W. Winston and R.E. Hayes. Optoelectronic device simulation of Bragg reflectors and their influence on surface-emitting laser characteristics. *IEEE J. Quantum Electron.*, 34:707–715, 1998.

[86] https://simwindows.wixsite.com/simwindows.

[87] J. Kong, Z. Yu, and Z. Yang. Stable and fast simulation of semiconductor lasers. In M. Osinski, P. Blood, and A. Ishibashi, editors, *Physics and Simulation of Optoelectronic Devices VI*, volume 3283 of *Proc. SPIE*, pages 396–403, 1998.

[88] U. Bandelow, H. Gajewski, and H.-C. Kaiser. Modeling combined effects of carrier injection, photon dynamics and heating in strained multi quantum well lasers. In R.H. Binder, P. Blood, and M. Osinski, editors, *Physics and Simulation of Optoelectronic Devices VIII*, volume 3944 of *Proc. SPIE*, pages 301–310, 2000.

[89] U. Bandelow, R. Huenlich, and T. Koprucki. Simulation of static and dynamic properties of edge-emitting multi-quantum-well lasers. *IEEE J. Select. Topics Quantum Electron.*, 9:798–806, 2003.

[90] http://www.cfd-research.com/.

[91] http://www.wias-berlin.de/software/tesca/index.html.en.

[92] B.P. Riely, J.J. Liu, H. Shen, G. Dang, W.H. Chang, Y. Jiang, Z. Sikorski, T. Czyszczanowski, and A.J. Przekwas. Multiphysics modeling of intracavity-contacted vcsel. In M. Osinski, H. Amano, and P. Blood, editors, *Physics and Simulation of Optoelectronic Devices XI, 27 January 2003, Can Jose, California*, volume 4986 of *Proceedings, Society of Photo-Optical Instrumentation Engineers (SPIE) Conference*. SPIE-The International Society for Optical Engineering, 2003.

[93] A.I. Fedoseyev, A. Przekwas, M. Turowski, and M.S. Wartak. Robust computational models of quantum transport in electronic devices. *Journal of Computational Electronics*, 3:231–234, 2004.

[94] http://www.rsoftdesign.com/products/component_design/LaserMOD/.

[95] http://www.silvaco.com/index.html.

[96] http://www.silvaco.com/index.html.

[97] http://www.silvaco.com/products/vwf/atlas/vcsels/vcsels.html.

[98] http://www.nextnano.de.

[99] http://www.photond.com.

[100] P. Vankwikelberge, G. Morthier, and R. Baets. CLADISS-a longitudinal multimode model for the analysis of thestatic, dynamic, and stochastic behavior of diode lasers withdistributed feedback. *IEEE J. Quantum Electron.*, 26:1728–1741, 1990.

[101] http://www.semitech.us.

[102] K.A. Bulashevich, I.Yu. Evstratov, and S.Yu. Karpov. Hybrid ZnO/III-nitride light-emitting diodes: modelling analysis of operation. *phys. stat. solidi (a)*, 204:241–245, 2007.

[103] http://www.synopsys.com.

[104] http://www.ise.ch.

[105] http://www.nusod.org/nusod04/ISE-NUSOD-2004.pdf.

[106] http://www.nusod.org/.

[107] J. Piprek, Cun-Zheng Ning, J. Wunsche, and H. Siu-Fung Yu, editors. *NUSOD'02. Introduction to the issue on optoelectronic device simulation*, volume 9 of *IEEE J. Select. Topics Quantum Electron.* May-June 2003.

[108] J. Piprek, E. Larkins, and S.-F. Yu, editors. *NUSOD'06. Special Issue on Numerical Simulation of Optoelectronic Devices*, volume 38 of *Optical and Quantum Electronics.* 2006.

[109] http://www.iwce.org/.

[110] D. Yevick, editor. *Physics and Simulation of Optoelectronic Devices, 25-26 March 1992, Somerset, New Jersey, USA*, volume 1679 of *Proceedings, Society of Photo-Optical Instrumentation Engineers (SPIE) Conference.* SPIE-The International Society for Optical Engineering, 1992.

[111] W.W. (Weng W.) Chow and Marek Osinski, editors. *Physics and Simulation of Optoelectronic Devices II, 24-26 January 1994, Los Angeles, California*, volume 2146 of *Proceedings, Society of Photo-Optical Instrumentation Engineers (SPIE) Conference*. SPIE-The International Society for Optical Engineering, 1994.

[112] Marek Osinski and W.W. Chow, editors. *Physics and Simulation of Optoelectronic Devices III, 6-9 February 1995, San Jose, California*, volume 2399 of *Proceedings, Society of Photo-Optical Instrumentation Engineers (SPIE) Conference*. SPIE-The International Society for Optical Engineering, 1995.

[113] Marek Osinski and W.W. (Weng W.) Chow, editors. *Physics and Simulation of Optoelectronic Devices IV, 29 January-2 February 1996, San Jose, California*, volume 2693 of *Proceedings, Society of Photo-Optical Instrumentation Engineers (SPIE) Conference*. SPIE-The International Society for Optical Engineering, 1996.

[114] M. Osinski and W.W. Chow, editors. *Physics and Simulation of Optoelectronic Devices V*, volume 2994 of *Proceedings, Society of Photo-Optical Instrumentation Engineers (SPIE) Conference*. SPIE-The International Society for Optical Engineering, 1997.

[115] P. (Peter) Blood, Marek Osinski, and Akira Ishibashi, editors. *Physics and Simulation of Optoelectronic Devices VI, 26-30 January, 1998, San Jose, California*, volume 1679 of *Proceedings, Society of Photo-Optical Instrumentation Engineers (SPIE) Conference*. SPIE-The International Society for Optical Engineering, 1998.

[116] Marek Osinski, Akira Ishibashi, and P. (Peter) Blood, editors. *Physics and Simulation of Optoelectronic Devices VII, 25-29 January, 1999, San Jose, California*, volume 3625 of *Proceedings, Society of Photo-Optical Instrumentation Engineers (SPIE) Conference*. SPIE-The International Society for Optical Engineering, 1999.

[117] Marek Osinski, P. (Peter) Blood, and Rolf H. Binder, editors. *Physics and Simulation of Optoelectronic Devices VIII, 24-28 January 2000, San Jose, USA*, volume 3944 of *Proceedings, Society of Photo-Optical Instrumentation Engineers (SPIE) Conference*. SPIE-The International Society for Optical Engineering, 2000.

[118] Marek Osinski, Yasuhiko Arakawa, and P. (Peter) Blood, editors. *Physics and Simulation of Optoelectronic Devices IX, 22-26 January 2001, San Jose, USA*, volume 4283 of *Proceedings, Society of Photo-Optical Instrumentation Engineers (SPIE) Conference*. SPIE-The International Society for Optical Engineering, 2001.

[119] Marek Osinski, Yasuhiko Arakawa, and P. (Peter) Blood, editors. *Physics and Simulation of Optoelectronic Devices X, 21-25 January 2002, San Jose, USA*, volume 4646 of *Proceedings, Society of Photo-Optical Instrumentation Engineers (SPIE) Conference*. SPIE-The International Society for Optical Engineering, 2002.

[120] M. Osinski, H. Amano, and P. Blood, editors. *Physics and Simulation of Optoelectronic Devices XI, January 2003, San Jose, USA*, volume 4986 of *Proceedings, Society of Photo-Optical Instrumentation Engineers (SPIE) Conference*. SPIE-The International Society for Optical Engineering, 2003.

[121] Marek Osinski, Hiroshi Amano, and Fritz Henneberger, editors. *Physics and Simulation of Optoelectronic Devices XII, January 2004, San Jose, USA*, volume 5349 of *Proceedings, Society of Photo-Optical Instrumentation Engineers (SPIE) Conference*. SPIE-The International Society for Optical Engineering, 2004.

[122] M. Osinski, F. Henneberger, and H. Amano, editors. *Physics and Simulation of Optoelectronic Devices XIII, January 2004, San Jose, USA*, volume 5722 of *The Society of Photo-Optical Instrumentation Engineers (SPIE) Conference*. SPIE-The International Society for Optical Engineering, April 2005.

[123] M. Osinski, F. Henneberger, and Y. Arakawa, editors. *Physics and Simulation of Optoelectronic Devices XIV, January 2006, San Jose, California, USA*, volume 6115 of *The Society of Photo-Optical Instrumentation Engineers (SPIE) Conference*. SPIE-The International Society for Optical Engineering, February 2006.

[124] M. Osinski, F. Henneberger, and Y. Arakawa, editors. *Physics and Simulation of Optoelectronic Devices XV, 22-25 January 2007, San Jose, California, USA*, volume 6468 of *Society of Photo-Optical Instrumentation Engineers (SPIE) Conference*. Bellingham, Wash. : SPIE, c2007., February 2007.

[125] Marek Osinski, Fritz Henneberger, and Keiichi Edamatsu, editors. volume 6889 of *Proceedings, Society of Photo-Optical Instrumentation Engineers (SPIE) Conference*. SPIE-The International Society for Optical Engineering, 2008.

[126] Marek Osinski, Bernd Witzigmann, Fritz Henneberger, and Yasuhiko Arakawa, editors. volume 7211 of *Proceedings, Society of Photo-Optical Instrumentation Engineers (SPIE) Conference*. SPIE-The International Society for Optical Engineering, 2009.

[127] Bernd Witzigmann, Fritz Henneberger, Yasuhiko Arakawa, and Marek Osinski, editors. volume 7597 of *Proceedings, Society of Photo-Optical*

Instrumentation Engineers (SPIE) Conference. SPIE-The International Society for Optical Engineering, 2010.

[128] Bernd Witzigmann, Fritz Henneberger, Yasuhiko Arakawa, and Alexandre Freundlich, editors. volume 7933 of *Proceedings, Society of Photo-Optical Instrumentation Engineers (SPIE) Conference.* SPIE-The International Society for Optical Engineering, 2011.

[129] Bernd Witzigmann, Marek Osinski, Fritz Henneberger, and Yasuhiko Arakawa, editors. volume 8255 of *Proceedings, Society of Photo-Optical Instrumentation Engineers (SPIE) Conference.* SPIE-The International Society for Optical Engineering, 2012.

[130] Bernd Witzigmann, Marek Osinski, Fritz Henneberger, and Yasuhiko Arakawa, editors. volume 8619 of *Proceedings, Society of Photo-Optical Instrumentation Engineers (SPIE) Conference.* SPIE-The International Society for Optical Engineering, 2013.

[131] Bernd Witzigmann, Marek Osinski, Fritz Henneberger, and Yasuhiko Arakawa, editors. volume 8980 of *Proceedings, Society of Photo-Optical Instrumentation Engineers (SPIE) Conference.* SPIE-The International Society for Optical Engineering, 2014.

[132] Bernd Witzigmann, Marek Osinski, Fritz Henneberger, and Yasuhiko Arakawa, editors. volume 9357 of *Proceedings, Society of Photo-Optical Instrumentation Engineers (SPIE) Conference.* SPIE-The International Society for Optical Engineering, 2015.

[133] Bernd Witzigmann, Marek Osinski, and Yasuhiko Arakawa, editors. volume 9742 of *Proceedings, Society of Photo-Optical Instrumentation Engineers (SPIE) Conference.* SPIE-The International Society for Optical Engineering, 2016.

[134] Bernd Witzigmann, Marek Osinski, and Yasuhiko Arakawa, editors. volume 10098 of *Proceedings, Society of Photo-Optical Instrumentation Engineers (SPIE) Conference.* SPIE-The International Society for Optical Engineering, 2017.

[135] Bernd Witzigmann, Marek Osinski, and Yasuhiko Arakawa, editors. volume 10526 of *Proceedings, Society of Photo-Optical Instrumentation Engineers (SPIE) Conference.* SPIE-The International Society for Optical Engineering, 2018.

[136] Bernd Witzigmann, Marek Osinski, and Yasuhiko Arakawa, editors. volume 10912 of *Proceedings, Society of Photo-Optical Instrumentation Engineers (SPIE) Conference.* SPIE-The International Society for Optical Engineering, 2019.

[137] Bernd Witzigmann, Marek Osinski, and Yasuhiko Arakawa, editors. volume 11274 of *Proceedings, Society of Photo-Optical Instrumentation Engineers (SPIE) Conference*. SPIE-The International Society for Optical Engineering, 2020.

[138] Bernd Witzigmann, Marek Osinski, and Yasuhiko Arakawa, editors. volume 11680 of *Proceedings, Society of Photo-Optical Instrumentation Engineers (SPIE) Conference*. SPIE-The International Society for Optical Engineering, 2021.

[139] Bernd Witzigmann, Marek Osinski, and Yasuhiko Arakawa, editors. volume 11995 of *Proceedings, Society of Photo-Optical Instrumentation Engineers (SPIE) Conference*. SPIE-The International Society for Optical Engineering, 2022.

[140] A. Mooradian, J.P. Donnelly, C.S. Harder, and K. Iga. Introduction to the issue on semiconductor lasers. *IEEE J. Select. Topics Quantum Electron.*, 7:93–95, 2001.

[141] M.R. Krames, H. Amano, J.J. Brown, and P.L. Heremans. Introduction to the issue on high-efficiency light-emitting diodes. *IEEE J. Select. Topics Quantum Electron.*, 8:185–188, 2002.

[142] J. Piprek, Cun-Zheng Ning, H. Wunsche, and Siu-Fung Yu. Introduction to the issue on optoelectronic device simulation. *IEEE J. Select. Topics Quantum Electron.*, 9:685–687, 2003.

[143] S. Arai, K.J. Ebeling, and Y. Yoshikuni. Introduction to the issue on semiconductor lasers. *IEEE J. Select. Topics Quantum Electron.*, 9:1111–1112, 2003.

[144] L.A. Coldren, A. Larsson, and F. Koyama. Introduction to the issue on semiconductor lasers. *IEEE J. Select. Topics Quantum Electron.*, 13:1043–1045, 2007.

Index

$n - P$ heterojunction, 76
$p - N$ heterojunction, 74
3-layer structure, 227

Absorption, 254
    in a two-level system, 4, 5
Acceptor concentration, 61
Active region, 10, 12
    in a VCSEL, 10
AlGaAs, 35
Arsenic, 27
Atomic orbitals, 21
Auger non-radiative recombination,
    87
Auger processes, 257

Band discontinuity, 307
Band edges, 123
Band gap energy, 80
Band offset, 307
Band structure, 21
Band structure parameters, 80
Bernoulli function, 165
Bias voltage, 71
Boltzmann distribution., 24
Boltzmann statistics, 42
Boron, 157
Boundary conditions, 139, 158

Carrier transport equations, 156
Charge neutrality, 29, 62
Cladding, 233
Compensated semiconductor, 30
computational electronics, 94
Conduction band, 8
Continuity equations, 42
Convergence, 111
Current-voltage characteristics, 1

Debye length, 64
Degenerate semiconductors, 25
Densities, 159
Depletion approximation, 66
Diamond structure, 19
Diffusion coefficient, 44
Direct semiconductors, 21
Dirichlet boundary conditions, 95
Discretization, 95
Distributed Bragg reflector (DBR),
    10, 11
Donor concentration, 61
Doping, 81
Double-heterostructure, 13
Drift-diffusion model, 41

Effective density of states, 81
Einstein relations, 42
Electric current, 170
Electric field, 234
Electrical equations, 156
Electrical variables, 161
Electromagnetic field, 13
Electron affinity, 80
Electron capture, 87
Electron concentration, 26
Electron emission, 87
Electron-hole pairs, 8
Electrons, 9
Emission
    spontaneous, 4, 5
    stimulated, 4, 5
Energy gap, 19
Equilibrium, 23
Excited state, 4

Fabry-Perot, 8

Fabry-Perot (FP)
   resonance conditions, 6
   resonant modes, 7
   resonator, 6
Far-field pattern, 14
Feedback, 2
Fermi energy, 23
Fermi level, 11
Fermi-Dirac distribution, 23
Field dependent mobility, 158
Field independent mobility, 157
Flatband conditions, 76
Forward bias, 70
Free-carrier absorption, 255

GaAs, 19
Gain, spectrum of semiconductor
   laser, 7
Gain spectra, 252
Generation-recombination processes,
   156
Germanium, 19
Gradient model, 55
Group IV atoms, 27
Group velocity, 87
Gummel's algorithm, 94

Hetero p-n junction, 74
Heterostructure, 30
Hole capture, 87
hole emission, 87
Holes, 9
Homostructures, 30

Impurities, 23
Impurity energy levels, 81
Indirect semiconductors, 21
InGaAsP, 311
Initial conditions, 111
Intervalence band absorption, 256
Intrinsic concentration, 99
Intrinsic semiconductor, 23

Laser diode
   distributed feedback laser
      (DFB), 10

in-plane laser, 9
vertical cavity surface-emitting
   laser (VCSEL), 9, 10
Local optical gain, 263
Losses, 266
Low-field mobility, 158

Matlab, 94
Maxwellian distribution function, 53
Mesh points, 159
Metal-semiconductor (MS) contacts,
   49
Metallurgical junction, 61
Mirror loss, 257
Mobility, 44, 83
Mobility models, 157
Multilayer waveguide, 229

n-region, 64
Non-radiative processes, 86
Nondegenerate semiconductors, 43

Ohmic contacts, 49, 51, 70
Optical cavity, 3
   Fabry-Perot, 6
Optical equations, 265
Optical parameters, 250
Oscillator, 2
Output power, 14

p-i-n diode, 73
p-n junction, 61
   double heterostructure, 12
   homogeneous (homojunction),
      11–12
p-region, 65
Peak gain, 253
Phosphorous, 157
Photon rate equation, 266
Physical constants, 79
Planck potentials, 42
Planck's radiation law, 4
Poisson equation, 64, 94
Poisson equation in equilibrium, 261
Poisson equation in non-equilibrium,
   261

Potential at contacts, 159
Profiles of electric field, 226
Propagation constant, 232
Propagation constants, 228
Propagation matrix, 230, 232
Pseudopotential method, 21
Pumping, 2, 3

Quantum well, 10, 38
Quasi-Fermi level, 11

Radiative processes, 86
Recombination parameters, 79
Recombination processes, 86
Refractive index, 226, 250
Relative dielectric constant, 138, 250
Reverse bias, 70

Saturation velocity, 158
Scaling, 166
Scharfetter-Gummel scheme, 162
Schottky barriers, 50
Schottky contacts, 49
Sellmeier formula, 251
Semiconductor laser, 1
Semiconductor transport equations,
    41
Semiconductors, 19
separate confinement
    heterostructures, 38
Shockley-Read-Hall recombination,
    86

Silicon, 19
Slab waveguide, 226
slab waveguides, 265
Spontaneous emission, 257
Spontaneous recombination, 87
Static permittivity, 268
Steady state, 165
Stimulated recombination, 87
Substrate, 233

TE modes, 227
Tetrahedral structure, 19
Thermionic model, 52
Threshold current, 14
Transitions, in a two-level system,
    4–5
Trial values, 160
Type I heterostructure, 35
Type II heterostructure, 36
Type III heterostructure, 36

Valence band, 8, 35
Vertical cavity surface emitting laser,
    9

Wave equation, 265
Wave function, 21
Waveguide, 230
    lossy, 234

Zincblende structure, 19

Printed in the United States
by Baker & Taylor Publisher Services